From field to lab
200 life science experiments for the amateur biologist

James D. Witherspoon, PhD

Professor of Biology
Grand Canyon University

Illustrated by the author and his wife,
Rebecca H. Witherspoon

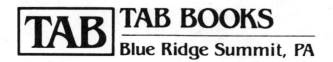
TAB BOOKS
Blue Ridge Summit, PA

764068

To my students in America, China, and Hungary

FIRST EDITION
FIRST PRINTING

© 1993 by **TAB Books**.
TAB Books is a division of McGraw-Hill, Inc.

Library of Congress Cataloging-in-Publication Data

Witherspoon, James D.
     From field to lab : 200 life science experiments for the amateur /
by James D. Witherspoon.
         p.   cm.
     Includes index.
     ISBN 0-8306-4103-3   ISBN 0-8306-4104-1 (pbk.)
     1. Physiology, Experimental—Laboratory manuals.   2. Biology,
Experimental—Laboratory manuals.   I. Title.
     QP42.W6   1992
     574′.078—dc20   92-32121
                                                                   CIP

Acquisitions Editor: Kim Tabor
Editor: Susan J. Bonthron
Supervising Editor: Joanne M. Slike
Direction of Production: Katherine G. Brown
Book Design: Jaclyn J. Boone
Cover Design and Illustration: Sandra Blair Design,
  Harrisburg, Pa.   TAB1

Contents

PART TWO
ANIMALS WITHOUT BACKBONES

Acknowledgments

*T*hank you Becky Witherspoon for helping me with the experiments, manuscript, and illustrations.

Thank you Kara and Tim Mohr and Jeff Lime for trying many of these experiments. I hope your students continue to enjoy doing them in the years ahead.

Thank you Reed Click, John Smoot, and Al Weir for helping me collect the data for Figures 33-1 and 33-2. Thank you International Biomedical for permission to use these published graphs.

This book is an outgrowth of an earlier book, *The Living Laboratory*, published by Doubleday & Co. in 1960. My wife and I wrote it while still graduate students. I thank the editors of Doubleday for accepting the work of young, unknown authors and the editors of Addison-Wesley, Harper & Row, and TAB Books for their help with later books. I particularly thank Kim Tabor, the Editor-in-Chief of TAB Books, for her enthusiastic support. Good editors and eager readers spur authors ahead.

Introduction

*F*or those who want to know nature, there is no better way than to pry off the lid and look inside. In the following chapters and experiments, you do just that. First you read to gain knowledge of each biological topic. Then you gather a few commonly available materials from home or school and begin one of the various field or laboratory investigations. In these you swim among fish, follow tracks in snow, measure reaction times of friends, and watch the heartbeat of a flea—to mention a few. Then, if you wish, you try "other activities" at the ends of the chapters. Some point to uncharted areas of biology in which lie the real thrills for beginner and professional alike.

Using this book, you will observe insects, fish, amphibians, reptiles, and other animals—but the emphasis is on humans. You contract and feel the actions of your muscles, check your reflexes, test methods of learning, measure your pulse while underwater, hold your breath before and during exercise, and so on.

This book is for middle- and high-school students, those who love to observe and experiment, and for amateur scientists of all ages. Some experiments require no equipment—only you and your imagination. Others require simple materials from your home, school, grocery, drugstore, or a biological supply house.

When observing or experimenting on live vertebrates, such as tadpoles or fish, imagine yourself in their place. Their nervous systems are less developed than yours, but they do have feelings. Treat animals with kindness and respect and if you remove them from their original home, put them back when you finish.

As you read, watch for occasional warnings in bold print, placed there to assure your safety. Alkaline solutions, preservatives, and other potentially harmful substances are occasionally used where a substitute, in the author's opinion, provides poor results. These agents are hazardous to experimenters somewhat as bleaches and antifreeze are hazardous. For experiments with warnings, use adult supervision—a parent or teacher—and follow the procedures as written. If you spill a hazardous substance, and particularly if you get any on your body, wash it off with lots of water. Though there is little chance for damage if the warnings are followed, neither the publisher nor the author can assume responsibility.

This book is for people who want to learn and have fun, to explore and experience. Some experiments are simple; others are challenging and require care to get accurate results. Take pride in your endeavors and learn these lessons well. You have a brilliant, rewarding future on earth or in space, a future well worth the effort that makes it possible.

Part I

Embryos, cells, and genes

Chapter **I**

The beginning
of life

On the yolk of every fertile chicken egg, there is a pinhead-sized, simple-looking white spot. If a hen sits on the egg, this spot becomes an animated baby chick in just three weeks. Here is a Cinderella-like miracle, a turning of mice into horses, and pumpkins into carriages. The cells of this spot contain the blueprint for life.

No matter how we turn the egg, the spot drifts upward because it is lighter than the yolk. This position brings the unborn *embryo* closer to the mother hen. The heat of her breast provokes little change at first—just the continued division of cells that began while the egg was inside her. But toward the end of the first day, the cells drift inward to start a spinal cord and brain. The head and tail ends differentiate. And in the second day, a beating heart and blood vessels appear. The vessels progress slowly over the yolk where they pick up nutrients.

As the days pass, more structures develop, including skin, bone, eyes, nerves, muscles, and other parts—all in miniature. The embryo begins to look like a chicken, and we recognize this by renaming it a *fetus*. Eight to ten days have passed since the hen began brooding.

Near the end of incubation, the fetus pokes its beak into an air cell at the blunt end of the egg, where it breathes. (Perhaps you have noticed this space when opening boiled eggs.) Then, ready to emerge, it pecks at the inside of the shell, cracking and breaking its way into a new world. Its three weeks of captivity have ended.

Chicken embryos are excellent subjects for study. You can easily obtain fertile eggs from hatcheries, and open them. But the miracle of the developing chick is repeated throughout the animal kingdom. Even humans begin as barely visible, fertilized eggs; of course, in them there is no yolk, white, or hard outer shell.

The human egg is fertilized in a *uterine tube* attached to the womb or *uterus*. The egg divides repeatedly, forming a mass of about 100 cells by the time it reaches and implants in the uterus. The implanted mass is the embryo. Like the chick embryo, it develops a brain, spinal cord, heart, limb buds, and other structures; but unlike the chick embryo, it sprouts an *umbilical cord*. The umbilical cord contains blood vessels that draw nourishment from the mother's uterus—the embryo's living incubator (FIG. 1-1).

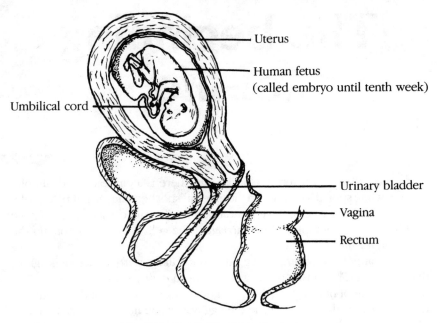

1-1   Four-month fetus in mother's uterus

The young embryos of all animals bear a marked resemblance to each other. Humans start to grow gills at one stage, as if to become fish, and sprout tails, as if to become monkeys. Later the gills and tails disappear, making the embryos look human. This sequence takes 8 to 10 weeks. Now the embryos are called fetuses, and they remain fetuses until birth.

Materials

- Chicken eggs incubated for four and eight days
- Hand lens or stereomicroscope
- Watch glass or plastic lid
- 1000-milliliter beaker or measuring cup
- Sodium chloride (table salt)
- Sharp-pointed scissors

OPENING AN INCUBATED EGG

From a hatchery or hen, obtain eggs that have been incubated for four and eight days. Use them within two or three hours or keep them at 33° to 39° C (91° to 102° F). Dissolve 9 grams of sodium chloride (table salt) in 1 liter of water in a large beaker or measuring cup. If you do not have a balance that weighs in grams, use 2 level teaspoons of salt in 1 quart of water. Heat the solution to 39° C (102° F), and keep it at or near this temperature throughout the experiment.

Start with the egg that was incubated for four days. With the sharp point of scissors, punch a small hole in the eggshell about a third down from one end, preferably the end that most recently pointed up (FIG. 1-2). The embryo floats upward. Insert the tip of the scissors into the hole, and cut carefully around the shell. Avoid the yolk and embryo that lie deeper than the egg white.

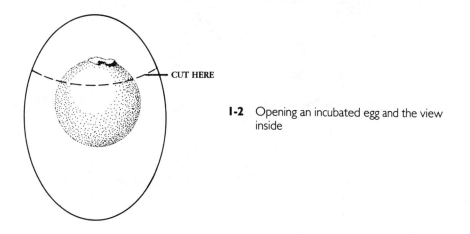

1-2 Opening an incubated egg and the view inside

When finished, hold the egg at the surface of the salty water, and open it. Spill the contents gently into the solution. The salt concentration of the solution is about the same as that of the egg and, therefore, adapted to the needs of the embryo. Rotate the yolk, if necessary, to find the embryo. It will appear as a definitely formed, weirdly arranged animal about half an inch long (FIG. 1-3). The size and extent of embryonic development vary.

OBSERVING A FOUR-DAY-OLD EMBRYO

When you first see the four-day embryo, look for its beating heart! (The heart may continue pumping for minutes or hours.) Then with scissors, gently cut the embryonic attachments and the embryo free from the yolk. Slide a clean watch glass or plastic lid under the embryo, removing the embryo and some salt solution for a closer look. Use a hand lens or low-power stereomicroscope for magnification.

Watch the blood as it pulses from the two-chambered heart. Notice the network of blood vessels. The heart later becomes four-chambered, and the

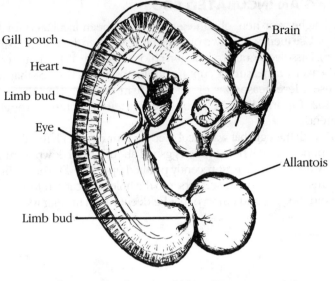

Gill pouch

Heart

Limb bud

Eye

Brain

Allantois

Limb bud

1-3   Four-day chick embryo

vessels sprout many branches. Examine the head. It has expanded to hold the brain, out of which grow large eyes. Observe segments of the vertebral column—the future backbone—and of the spinal cord that lies inside it. Look for limb buds at the site of future legs and wings. Between the leg buds, a pouch—the allantois—projects outward from the intestine. Nitrogen-containing wastes are deposited in this pouch. Look also for gill pouches, which are ridges separated by grooves at the front of the neck. Although they resemble the gill pouches of fish embryos, those of birds are later resorbed; they never become working gills.

OBSERVING AN EIGHT-DAY-OLD EMBRYO

Remove the embryo from an eight-day egg, and transfer it to a watch glass or plastic lid containing salt solution. The appearance now is more like that of an adult chicken (FIG. 1-4). The limb buds are much further developed. Probably you can see fingers (later to be wings) and toes. Look for a primitive beak and for the opening of an ear behind the eyes. The bumpy skin, resembling that of a plucked hen, shows where feathers will bloom.

OTHER ACTIVITIES

You have seen two greatly different stages of embryonic development. If you wish, you can study the day-to-day changes to better see how individual organs evolve and in what sequence. To do this, obtain eggs of one through eight days of age.

You might also incubate your own eggs. For this purpose, buy fertile eggs at a hatchery, and keep them in a semihumid container heated to 38° to 40° C

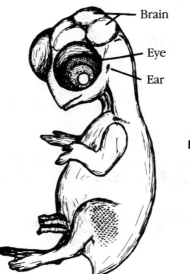

Brain

Eye

Ear

I-4 Eight-day embryo

(100° to 104° F). The incubator can be of many designs. I once successfully used a bun warmer over a pilot light. You might prefer to use a Styrofoam box in which you cut a hole for a plastic window. Heat the box with a small light bulb, trying different bulbs until you get a suitable temperature. To be sure the temperature and humidity are correct, put the bulb of a thermometer at egg-level and a small pan of water nearby.

Whatever your incubator, turn the eggs three or four times daily to prevent embryonic membranes from adhering to the shells. To see the embryos develop, open new eggs once or twice daily.

Alternatively, tape plastic wrap over a small opening you make in each of several eggs. Keep the opening as clean as possible to prevent contamination and death. Then incubate the eggs. Look through the transparent plastic to see embryonic growth hour by hour. You can watch the complete embryonic progression of an organ, such as the heart, or the growth of eyes from the brain.

Perhaps you would prefer to study another animal. If so, purchase fertilized frog eggs from a biological supply house (Appendix B), and follow the development of the embryos into tadpoles and frogs. Use the directions given by the supplier. To see the effects of temperature on development, keep some of the embryos at 10° to 15° C (50° to 59° F) and others at 20° to 25° C (68° to 77° F).

Chapter 2

Protozoa and other cells

*U*ntil the 1600s, people saw distant objects, such as stars, and small objects, such as lice, with only their eyes or eyeglasses. Then our universe expanded because of two inventions: the telescope and the microscope. Galileo Galilei turned his homemade telescope to the sky, discovering countless stars and the moons of Jupiter. He confirmed that the sun, not the earth, is the center of the solar system. Soon after, Antoni van Leeuwenhoek turned his homemade microscope to drops of water, discovering countless small creatures—mostly protozoa and bacteria. In one of these drops, wrote Leeuwenhoek:

> I now saw very plainly . . . little eels, or worms, lying all huddled together and wriggling; just as if you saw, with the naked eye, a whole tubful of very little eels and water, with the eels a-squirming among one another. . . . No more pleasant sight has ever yet come before my eye than these many thousands of living creatures, seen all alive in a little drop of water. . . .[1]

Leeuwenhoek's hand-held microscopes had single lenses, which he ground himself, but they magnified the image up to 270 times. Thereby he saw even the structures within protozoa, revealing the complexity of their bodies.

Leeuwenhoek was a merchant by trade, not a professional biologist. He was an amateur like you, but in a lifetime of peeking through lenses, he saw the development of ants, the spinning apparatus of spiders, the circulation of blood through capillaries, starch granules in plants, the structures of skin, teeth, muscle, and eyes, and other objects.

Leeuwenhoek's contemporary, Robert Hooke, made compound microscopes—those having two lenses—such as we use today. With these microscopes, Hooke saw that cork contains "cells" like the rooms of a monastery. We still use his word *cells* for the boxlike units of protoplasm that make up plants and animals.

If only Hooke and Leeuwenhoek could return! They would be amazed by the complexity of electron microscopes, the growth of cells outside the body, the separation and study of cellular components, and the splicing of genes from animals into bacteria. They would be thrilled by our microscopic progress.

Materials

- Microscope
- Glass slide
- Hair or other small object

USING A MICROSCOPE

Today you will collect protozoa and observe them through a microscope, as Leeuwenhoek did over 300 years ago. To prepare, look first at your microscope and at FIG. 2-1. The instruments each have two lenses, one above the other, allowing the upper lens to magnify the image of the lower. The one at the top of the tube is called an *ocular lens* and the one at the bottom an *objective lens*. Often there are two or more objective lenses on a rotating *nosepiece*, each lens with different powers of enlargement. To change powers, turn the nosepiece until the desired objective lens is aligned with the ocular lens.

To use the microscope, put a hair or other small object in the center of a rectangular glass *slide*, and clamp the slide to the *stage*. Rotate the *reflecting mirror* or turn on the light under the stage to direct strong light on the object. Adjust the *diaphragm*, between the light source and the stage, to reduce this light to a desired level. Now turn the *coarse adjustment knob* cautiously to bring your lowest-power objective lens downward until it is two or three millimeters (about one-eighth inch) from the slide. Then draw the objective lens slowly upward to bring the hair or other object into focus. Switch to a higher objective lens, if desired, being careful that the lens and slide do not touch. Revolve the *fine adjustment knob* to explore details.

You are looking at an object enlarged one hundred to several hundred times its normal size. Imagine the excitement of seventeenth-century scientists who first saw what you see now.

Materials

- Microscope
- Slide and cover glass
- Medicine dropper
- Water samples from ponds and ditches

OBSERVING PROTOZOA AND OTHER SMALL ORGANISMS

Obtain samples of water and decaying pond plants from local ponds and neighborhood ditches. Include some of the surface scum. Examine the

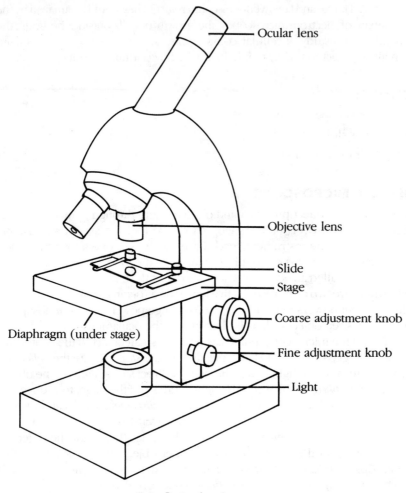

Ocular lens

Objective lens

Slide

Stage

Coarse adjustment knob

Fine adjustment knob

Light

Diaphragm (under stage)

2-1   Parts of a microscope

samples at once or, better yet, allow them to stand in open containers for one or two days. This provides time for the protozoa to reproduce. For best results, add small pieces of boiled lettuce to the water. Bacteria feed on the lettuce; protozoa feed on the bacteria.

Protozoa are single-celled animals, but they are not simple in structure. They contain many *organelles*—little organs, such as nuclei and vacuoles—inside them. Each cell provides its own respiration, digestion, excretion, and so on.

When you are ready, study the diagrams of protozoa and of microscopic, multicellular organisms—gastrotrichs and rotifers—that might also be in the water (FIGS. 2-2, 2-3, and 2-4). If available, get a book on microscopic plants and animals from your local library. Then with a medicine dropper, get a drop or two of water from the bottom of your pond-water container. Place this on a

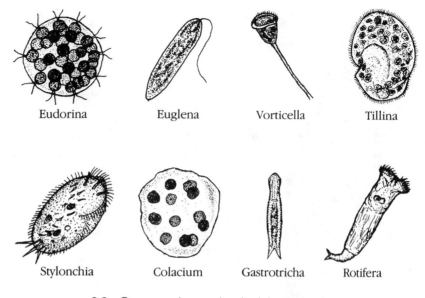

Eudorina   Euglena   Vorticella   Tillina

Stylonchia   Colacium   Gastrotricha   Rotifera

2-2   Common microscopic animals in pond water

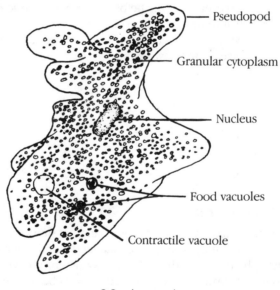

Pseudopod

Granular cytoplasm

Nucleus

Food vacuoles

Contractile vacuole

2-3   An amoeba

slide and lower a cover glass gently upon it, excluding all air bubbles. You can do this best by touching one edge of the cover glass to the slide, then slowly lowering the remainder of the glass onto the water and slide. Look for and identify the swimming, darting, creeping protozoa and other small organisms under low and high magnification. Reduce the light from the mirror or substage lamp to see details. Perhaps you will agree with Leeuwenhoek, "No

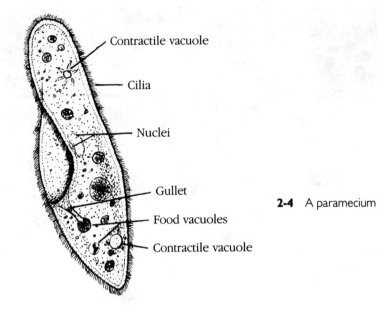

Contractile vacuole

Cilia

Nuclei

Gullet

2-4   A paramecium

Food vacuoles

Contractile vacuole

more pleasant sight has ever yet come before my eye than these many . . . living creatures, seen all alive in a little drop of water. . . ."

When finished with the first sample of water, obtain a second near the surface. Are other species in the new sample?

Materials

- Microscope
- Slide and cover glass
- Medicine dropper
- Water containing amoebas

OBSERVING AMOEBAS

Amoebas are single-celled transparent protozoans that slowly and continually change shape (FIG. 2-3). If they are in your container of pond water, you will find them mainly at the bottom. Put a drop of water containing them on a slide, and cover the drop with a cover glass. If there are no amoebas in the sample, you can purchase them from a biological supply house (Appendix B).

Watch the movements of the amoebas for a few minutes. Sketch one at different time intervals, and record the time of each drawing.

Like other cells, an amoeba has two parts: a nucleus and cytoplasm. The *nucleus* is the slightly darker, centrally located mass, perhaps difficult to distinguish. You might need to adjust the diaphragm under the stage, reducing the light, to see the nucleus and other small parts (FIG. 2-1). The nucleus contains DNA, the genetic substance that directs growth and reproduction. The *cytoplasm* is the transparent, granular material lying between the nucleus and the outermost *cell membrane*.

Within the cytoplasm are *vacuoles*—small, spherical containers that hold either food or water. Identify the empty-looking, water-holding *contractile vacuoles*. These pump excess water from the cell. Also identify *food vacuoles*, if present. *Pseudopods* of streaming cytoplasm engulf small life upon which the amoeba feeds, forming food vacuoles in which the organisms are digested.

Respiration and excretion are simple processes in amoebas and other protozoa. They absorb oxygen from the water and excrete carbon dioxide and other wastes into the water. Being small, they have no need for hearts, lungs, and kidneys. Gases and wastes simply diffuse.

Materials

- Microscope
- Slide and cover glass
- Medicine dropper
- Yeast
- Water containing paramecia

OBSERVING PARAMECIA

Place a drop of water containing paramecia on a slide, and cover the drop with a cover glass (FIG. 2-4). If there are no paramecia in your sample of pond water, you can purchase them from a supply house (Appendix B).

Observe the dashing, whirling, swimming motion of these slipper-shaped protozoans. Watch them back up as they meet obstructions, then attempt to find a clear path. Reduce the light of your microscope by partially closing its diaphragm. This enables you to see *cilia*—tiny, hairlike processes that are barely visible—covering the entire surface of each paramecium. Paramecia use cilia, like the oars of boats, to propel themselves through water.

Add to the paramecia a drop of yeast dissolved in water. Watch how the cilia draw yeast cells down the *gullet* to form a food vacuole. After a few moments, the first vacuole breaks off and another forms.

Locate nuclei, cytoplasm, and contractile vacuoles within the paramecia, and cell membranes around them. Paramecia have two or more faint nuclei and two contractile vacuoles. Allow the slide and paramecia partially to dry. Dehydration alters the shapes of paramecia but makes it easier to see their contractile vacuoles. Watch as these transparent, water-removing globes fill and empty. Look also for faint, radiating canals at the sides of the vacuoles. Water passes through the canals to the vacuoles.

Materials

- Microscope
- Flat-edged toothpick

- Slide and cover glass
- Tincture of iodine

OBSERVING CHEEK CELLS

Our bodies contain cells resembling those of protozoa. To see human cells, *gently* scrape the inner lining of your cheek with the flat edge of a toothpick. Scrape only *your* cheek, not that of another person, just as you would pick or brush only your teeth, to avoid the possibility of spreading infection.

Place the cheek scrapings in a drop of water on a slide, then discard the used toothpick in a trash bag. Cover the drop with a cover glass. Now look at the cells through a microscope. You will see flattened, transparent *epithelial cells* with small spherical nuclei in their centers. The edges of the cells may be folded (FIG. 2-5). Epithelial cells of different shapes line body cavities and tubes, and form the outer layer of the skin.

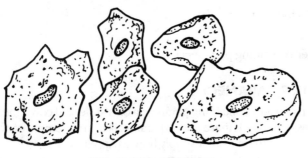

2-5   Cells from the cheek

Touch a minute drop of iodine to the edge of the cover glass. The color will pass under and stain the cells, especially the nuclei. Biologists often stain slides to enhance detail or to make previously imperceptible parts visible.

OTHER ACTIVITIES

Different cells combine to form different parts of the body. You can see these combinations by examining prepared slides of each, such as those you will find in the biology laboratories of high schools, colleges, and universities. Start, for example, with the villi of the small intestine, the fingerlike projections through which nutrients are absorbed. Continue with the alveoli of the lungs, the air sacs through which oxygen is transferred to the blood. Then look at the pancreas, a gland in which some cells secrete insulin and others secrete digestive enzymes. You will find descriptions of the intestine, lungs, pancreas, and other structures in textbooks of histology.

You can also make permanent mounts of microscopic objects, using methods described in histology manuals. The simplest procedure is to place small objects—such as hair, feathers, fish scales, or insect parts—in xylol. (See Appendix B for suppliers.) Begin by putting a circle of balsam on a microscope slide. Place a drop of xylol in the circle and the specimen in the xylol. Seal the circle by gently lowering a cover glass upon it, excluding air bubbles. Then imitating Leeuwenhoek, peer through a microscope at the mount for a "pleasant sight."

Endnotes

1. Leeuwenhoek, Antoni van. Quoted by Daniel J. Boorstin in *The Discoverers*, pp. 330–331. New York: Random House, 1983.

2. Leeuwenhoek, quoted by Boorstin, p. 331.

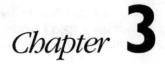

Movement of molecules

The cells of the body are like the buildings of a city. The different goods that workers bring into and out of the buildings give each building its identity. Likewise, the different materials that move into and out of cells give cells their identity. For these reasons, shoe stores differ from grocery stores, and cells of the liver differ from those of the pancreas. Yet all work together to make a viable organism.

Similarly, the walls and doors of buildings are like the membranes and pores of cells. The membranes maintain differences between the environments inside and outside the cells, and the pores admit and expel substances to maintain these differences. Molecules and ions pass through cell membranes, as when glucose goes from the intestine into the blood, or oxygen moves from the lungs into the blood.[1]

There are several methods of cellular transport; *diffusion* is one of them. In diffusion, molecules, ions, and other particles bounce about, moving from areas of greater concentration to areas of lesser concentration. If you open a bottle of perfume, for example, the molecules of the perfume diffuse out of the bottle, bouncing this way and that as they bump into other molecules, and soon reach your nose. Environmental heat provides the energy that moves the molecules.

Diffusion of perfume or other substances continues even when cell membranes intervene, as in the nose. Fat-soluble molecules pass because the membranes contain fat. Water-soluble molecules and ions pass, if they are small enough, because the membranes contain pores.

Another method of transport is *osmosis*. In osmosis, water diffuses through cell membranes or similar membranes from the side where it is more concentrated (more water molecules per unit of volume) to the side where it is less concentrated. In living systems, water is less concentrated where there are proteins or other particles in the water to displace it. For example, water moves by osmosis from the watery fluid outside capillaries into the bloody fluid inside capillaries because there is less water and more protein in the blood.

A third method of transport is *filtration*. In filtration, a liquid passes under pressure through a porous membrane. The key word is pressure. The pressure of blood, for example, forces small molecules, such as water and glucose, outward through the porous membranes of capillary blood vessels.

A fourth method of transport is *active transport*. In active transport, metabolic energy moves molecules and ions through cell membranes, often from sites of lesser concentration to sites of greater concentration. In the intestine, for example, glucose molecules are actively transported from the intestine into the blood, even when glucose is more concentrated in the blood. Active transport allows us to get the entire amount of glucose and other nutrients into the blood.

Materials

- Deep plastic tray or similar container
- Short barrier of plastic or other material
- 20 to 40 balls of Styrofoam or other material

MODEL FOR DIFFUSION AND ACTIVE TRANSPORT

Get a deep, preferably transparent plastic tray in which to place Styrofoam balls, rubber balls, pingpong balls, or marbles to represent molecules. Half of the balls should be one color and half another. For example, if you buy white Styrofoam balls, paint half of them black. Place a barrier that is ½ to 1 centimeter high (about ⅓ inch) across the center of the tray, dividing it into two sections. Place balls of one color on one side and of the other color on the other side (FIG. 3-1).

To demonstrate diffusion, shake the tray. As you do so, the balls will bounce and bump into each other. You are the source of energy for this movement, as heat is the source of energy for molecular movement. As you shake the tray, some of the white balls will bounce across the barrier to mix with the black balls and vice versa. The balls move from the side of higher concentration to that of lower concentration, as do molecules. Eventually the two colors become equally distributed.

To demonstrate active transport, shake the tray as a friend grabs the white balls that bounce to the black side and returns them to the white side. Your friend provides energy to move the balls in a specific direction, somewhat as your body provides energy to move molecules and ions. In the body, active transport often moves molecules and ions from sites of lesser to greater concentration.

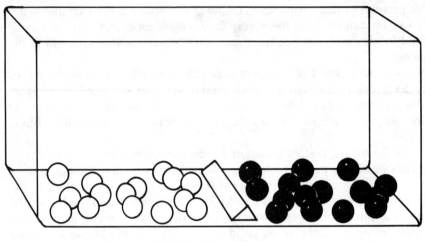

3-1   Model for diffusion

Materials

- Microscope
- Slide
- Cover glass
- Whole milk

EFFECTS OF MOLECULAR MOVEMENT IN MILK

Fast-moving molecules of a fluid bump other particles suspended within it. The bumping causes an erratic motion of the particles called *Brownian movement*.

You cannot directly watch the movements of molecules through ordinary microscopes because the magnification is too weak. But you can observe the effects of molecular movement on larger particles, such as the fat globules in milk. The fast- moving molecules collide with the globules, causing them to bounce short distances—far enough to be seen with a microscope.

Put a drop of whole milk (three to four percent fat) on a slide and lower a cover glass upon it. Wait a few minutes for the fluid to settle, then observe it under the highest power of your microscope. For best vision, reduce the light shining through the milk. Focus to see the jostling of fat globules by rapidly moving molecules.

Materials

- Perfume and paper towel or an orange
- A room with several people

DIFFUSION THROUGH A GAS

Do this experiment in a room where people stand or sit in different locations. When ready, soak a paper towel or tissue with cheap perfume, and unfold it to expose the largest possible surface. Have the people in the room tell when they each smell the odor. Alternatively, peel an orange and have people tell when they smell it. (If you use the orange, you can eat the experiment when finished.)

The odorous molecules diffuse through the air (a gas), bouncing from this molecule to that, until they penetrate the entire room. They are most concentrated near the source.

Materials

- Food color
- White cup filled with water

DIFFUSION THROUGH A LIQUID

Fill a white cup with water, and gently place one drop of food color in the water. Look at the cup every 5 to 10 minutes to see the dye slowly spread throughout the water. The dye continues to diffuse until it is evenly distributed.

Materials

- Food colors
- Knox gelatin in a bowl

DIFFUSION THROUGH A GEL

Prepare Knox uncolored gelatin as directed on the package. You can buy this at a grocery. When the gelatin has solidified, place one drop each of several food colors on it, keeping the drops several centimeters apart. Check the positions of the colors every half hour for several hours.

Though the gel appears solid, it is mainly water. Indeed, a single teaspoon of Knox gelatin gels up to two cups of water. The molecules of food colors gradually diffuse away from their centers of concentration.

Materials

- Dialysis tubing
- String
- Plastic pipette
- Dark molasses
- Bottle with two-hole rubber or cork stopper to fit

OSMOSIS

Stick a plastic pipet through one of the holes of a two-hole rubber or cork stopper (FIG. 3-2). Then tightly tie one end of a finger-length dialysis tube with string. Fold up the tied end of the tube, and tie it again to make the tube

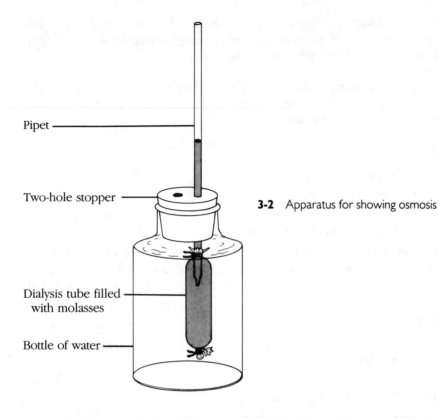

Pipet

Two-hole stopper

Dialysis tube filled
  with molasses

Bottle of water

3-2   Apparatus for showing osmosis

watertight. Fill the pouch with dark molasses. Next tie the open end of the
pouch tightly around the end of the pipette where it projects through the
stopper. Stick the stopper, pipet, and pouch of molasses into a bottle of water.
Position the top of the molasses and the top of the water at the same level.

The dialysis tube is permeable to water but not to the sugar in molasses.
Because there are more water molecules in water than in molasses, the net
movement of water is toward the molasses. This kind of diffusion is called
osmosis because it involves the movement of water through a semipermeable
membrane. What happens to the level of molasses in the pipet as water moves
into the pouch by osmosis? Check the rising fluid every 10 minutes for one or
more hours.

Materials

- Filter paper or paper towel
- Bottle or glass
- Funnel (optional)
- Paper cup
- Food color
- Powdered charcoal

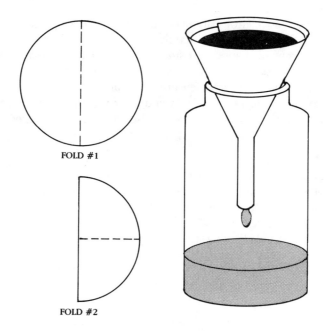

FOLD #1

FOLD #2

3-3 Folded paper and apparatus for filtration

FILTRATION

Fold a filter paper in half, and fold the half in half again to make a cone to insert into a funnel (FIG. 3-3). Set the funnel with its filter paper into the mouth of a bottle or glass. If you do not have the filter paper and funnel, simply cut and fold a circular piece of paper towel, and place it in the mouth of a glass. Thoroughly mix about one-half teaspoon each of food color and powdered charcoal in a paper cup of water. Pour the mixture into the funnel.

The pressure of water forces water and small particles through the pores of the filter paper. The pressure is highest when the funnel is fullest. What effect does this have on the rate of filtration at the beginning and end of the filtration? Which substance filters through and which remains behind? You can tell their locations by the color of the food color and the black of the charcoal. Which substance contains larger particles?

Your separation of large from small particles resembles that occurring in the capillaries of the kidneys. Small molecules and ions filter into the kidney tubules, but large molecules and blood cells remain in the bloodstream.

OTHER ACTIVITIES

Amoebas and paramecia (chapter 2) have salt and protein in their bodies, but the fresh water around them has none. By osmosis, therefore, some of the fresh water enters their salty bodies. If the water continues to enter and stay, they swell and perhaps burst, but it does not stay. Instead, contractile vacuoles (FIGS. 2-3 and 2-4) pump it out. The vacuoles fill with water, then contract to expel it.

Put some amoebas or paramecia under a microscope to observe the operation of their contractile vacuoles. How frequently do they fill and discharge? What would happen to osmosis if you added a little salt to the water outside the protozoans? Would water still enter their bodies? Would their vacuoles still discharge? Touch a few grains of salt to the edge of the water near the protozoa. What happens to them and their vacuoles as the salt diffuses around them? Can you devise a more precise method to test this effect?

Endnotes

1. Molecules, composed of one or more atoms, are the smallest particles of a substance that retain all its properties. Ions, for comparison, are atoms or groups of atoms that carry an electric charge. For example, sodium chloride (NaCl) is a molecule of salt. When it is placed in water, it splits into two ions, sodium (Na+) and chloride (Cl−).

Chapter **4**

Heredity in fruit flies

From the first raising of orphan cubs and planting of wild seed, our ancestors tried to improve upon nature. Breeders saw that offspring resembled parents, so they chose superior stock for mating. Yet the principles of genetics—the study of inheritance—were unknown. Improvement was slow.

Now that we better understand genetics, improvement is fast. By selecting better parents, animal breeders give us more eggs, milk, and meat at less cost. Plant breeders give us hybrid corn, resistance to frost and fungi in wheat, and faster growing, extremely productive rice. Such improvements are indirectly, or sometimes directly, the result of persistent studies by a vegetable-growing monk of the mid-1800s—Gregor Mendel.

Mendel was a favorite of his students at the high school in Austria and highly regarded by fellow monks; but biologists ignored him. Little did they know that Mendel's laws would someday become the backbone of genetics and that Mendel himself would achieve fame approaching that of his contemporary, Charles Darwin.

In a monastery garden, Mendel grew peas—common edible peas. He studied single, inheritable features—whether seeds were round or wrinkled, pods were green or yellow, plants were tall or short, and so on. Such characters are transmitted by *genes* located in the nuclei of reproductive cells.

Mendel carefully transferred the pollen from round-seeded pea plants to the pistils of wrinkled-seeded pea plants. (The pistil is the part of the flower that conducts pollen to the unfertilized seed.) When the progeny had grown, Mendel was surprised. His *first filial* (F_1 or offspring) generation contained only round peas. Roundness, therefore, is a *dominant* unit of heredity; wrinkled is *recessive*. Again Mendel transferred pollen, but this time he used F_1 plants, those that had both dominant and recessive genes. From hundreds of such crosses he found a consistent production of three round seeds for each wrinkled seed. The ratio was 3:1. The trait for wrinkled peas had skipped from the grandparents to the *second filial* (F_2) generation.

Mendel continued experimenting with other characters and other generations. He showed that tall plants were dominant to short plants and that green pods were dominant to yellow pods. He showed that two recessive parents—such as those that both had wrinkled seeds—bore only recessive offspring. Mendel helped find the rules by which breeders improve their stock and physicians predict the onset of disease.

Materials

- Stocks of wild and vestigial-winged fruit flies
- Four or more transparent vials of 50 to 200 milliliters capacity
- Culture medium
- Refrigerator with freezer
- Tuna can
- Plaster of Paris
- Hand lens or stereomicroscope

OBTAINING AND FEEDING FRUIT FLIES

Why use flies? Mainly because they are small and reproduce quickly. Fruit flies—more often called *Drosophila* by geneticists—are only two millimeters long. Because they are small, dozens or hundreds can be kept in a single vial. This provides large numbers to validate genetic principles. Also, they grow from egg to reproductive maturity in 10 to 12 days. This eliminates the months or years of waiting necessary in other animals to observe the results of a particular mating.

You probably have seen the brown, wild-type Drosophila flying around overripe fruit. You could trap and breed these, but in our experiment, you also need mutant, vestigial-winged flies. I suggest, therefore, that you purchase both types of flies, the culture medium in which they grow, and some vials in which to place the medium. You can get these items from suppliers listed in Appendix B.

In preparing new culture medium, use the directions from the supplier, putting 1 to 2 centimeters (½ inch) of the medium in each vial. Drop a tiny pinch of dried powdered yeast—just a few grains—on the medium before adding flies. To retard mold growth, use a freshly made medium for each generation of flies. (If you prefer to make rather than buy the culture medium, mix molasses with Cream of Wheat. This combination is easy to fix, yet solid enough that flies can be shaken from their bottles. For this purpose, use 11.5 milliliters (1 tablespoon) of molasses or Karo, 77.5 milliliters (⅓ cup) of water, and 10.3 grams (1 level tablespoon) of Cream of Wheat. Add about ⅔ of the water to the molasses and bring to the boiling point. Now pour in Cream of Wheat that has been mixed with the remaining ⅓ of the cold water. Cook and stir for a few minutes until the mixture thickens, then pour it into vials that have been sterilized by boiling. Add a tiny pinch of dried yeast to each vial, and you are ready for the flies.)

OBSERVING AND BREEDING FLIES

Examine the vials of Drosophila as soon as they come, noting the different life phases and sexes. Fruit flies have eggs, larvae, pupae, and adults. Use a hand lens and refer to FIG. 4-1. Find as many stages as possible. You can easily see wriggling larvae, immobile pupae, and adults, but the tiny, white eggs are all but invisible. Distinguish males from females. Male flies are smaller and have black-tipped abdomens. Females have fatter, more pointed abdomens. Compare Drosophila of the ordinary, long-winged, wild stock with those having vestigial wings. Vestigial wings are small and wrinkled, grounding their possessors.

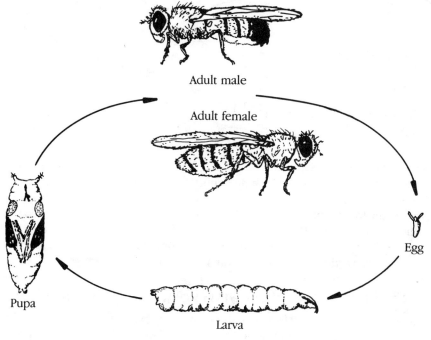

4-1 Life cycle of the fruit fly

You can begin your experiments at once or breed more flies. If you choose to breed pure stocks, transfer the supplier's wild flies to one vial of growth medium and vestigial-winged flies to another. If this is the first time you have handled flies, you might find it easier to immobilize them before transferring them. If so, put the vials on their sides in the refrigerator (*not* the freezer) for 30 minutes. The cold lowers the metabolism of the flies, making them less active. Then transfer the flies to new vials of culture medium, sprinkled with a few grains of yeast. The flies mate and a new generation of flies appears, identical with their parents. Any number of such generations breeds true unless a mutation occurs. This is very unlikely. The rate of gene mutations varies but is about one in one million.

IMMOBILIZING FLIES

Immobilize the flies when separating males from females. To prepare the apparatus for this, open and empty the contents of a circular can of tuna or chicken. Wash the empty can with soap and water. In a paper cup, thoroughly mix plaster of Paris with water to get a consistency about that of pancake batter. Pour the mixture into the circular can, filling it to the brim (FIG. 4-2). Allow 30 minutes or more for the plaster to harden, then place it in the freezer of a refrigerator for several hours.

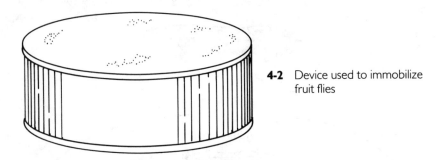

4-2   Device used to immobilize fruit flies

When you are ready to examine the flies, put their vial on its side in the refrigerator (not the freezer) for 30 minutes. After 25 minutes, remove the can of plaster from the freezer, letting it warm for 5 minutes. If you skip this step, you might freeze and kill the flies. Now sprinkle the flies from the refrigerated vial onto the plaster. The plaster should still be cold enough to keep the flies immobile for 10 to 20 minutes. This gives you time to sort males from females and mutant, vestigial-winged flies from the normal wild type.

MATING SHORT-WINGED WITH LONG-WINGED FLIES

Now you can test Mendel's ratios, those showing that heredity is directed by specific units (genes) and not by "blending of bloods." Start by using virgin females—a problem, since flies begin mating about 12 hours after birth. To get virgins, first remove all adult flies from your vials, saving only the eggs, larvae, and pupae. You can keep the adults in a separate vial, if you wish, or place them in the freezer to die. Then, 8 to 10 hours after the pupae start hatching, separate the adult males from the females. By these means the females remain unfertilized.

Mate virgin, vestigial-winged females with wild-type males or vice versa. Do so by placing several males of one type with several females of another type in a vial. Prepare two or more vials. Label each vial with the date and type of mating. In about a week remove the parents to prevent them from being confused with the offspring. Eggs become larvae in less than one day. Larvae molt twice and become pupae in about eight days. Pupae become adults a few days later, a complete life cycle in less than two weeks. Examine these offspring, the F_1 generation (FIG. 4-3). There is not a short-winged fly in the bunch! All are wild type. We see from this that long wings are dominant and vestigial wings, recessive.

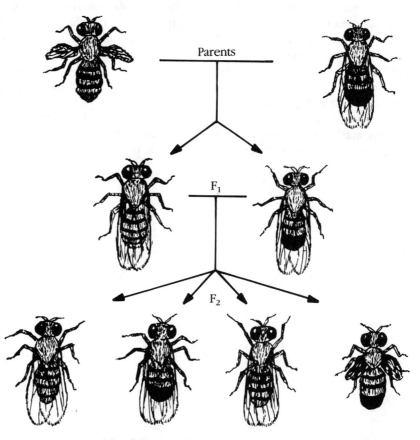

4-3 Offspring of wild type and vestigial flies

Put several F_1 pairs in two or three fresh vials and add labels. Virgins need not be used in this cross, since all F_1 flies have the same genetic constitution. Each carries one dominant and one recessive gene from its parents. Remove the F_1 flies after one week, during which they have mated and laid eggs. A few days later the F_2 flies start hatching, including some vestigials. Short wings skip one generation, but the recessive gene is still present. Immobilize and separate the flies into two classes: wild and vestigial. What is the ratio? Does it approximate Mendel's 3:1 ratio in peas?

OTHER ACTIVITIES

There are many Drosophila stocks for investigating principles of heredity. To observe shelves of a fruit-fly laboratory is an education in itself. Among the more distinctive varieties are those having ebony (black) bodies or sepia (dark brown) eyes. Because the wild type is dominant over each of these traits, you can substitute either stock for vestigial-winged flies in the preceding experiment.

As a further test, mate flies having ebony bodies with those having vestigial wings, but do not expect the offspring to look like their parents. From this cross you get only wild-type flies with ordinary brown bodies and long wings! Since both of these traits are dominant, they cover genes for black and vestigial. Now mate these F_1's to bring out recessives. Their offspring, the F_2's, will appear in the following ratio: nine wilds to three vestigials to three black bodies to one black body with vestigial wings.

For variety, plant corn seeds from a cross of green with albino corn. Such seeds are available from supply houses. The offspring will show a 3:1 ratio of green to albino in the seedling plants. Which gene is dominant?

Part 2

Animals without backbones

Chapter **5**

Hydra—the harmless serpent

*J*udging by numbers, single-celled organisms are the most successful form of life. Countless cells of this type are in every part of the world—in water and soil and as parasites in other animals.

But the single cells of many organisms group together for greater efficiency. *Volvox*, for example, is a colonial organization of thousands of cells—each connected to others by extensions of cytoplasm. Together they form a floating, green ball about the size of a pinhead. The coordinated colony swims as one body to capture food and reproduce.

Further united, but still primitive, are sponges and coelenterates—the latter including jellyfishes, sea anemones, and *hydra*. To picture hydra, imagine a short piece of string unraveled at one end (FIG. 5-1). When extended, this animal has the diameter of a thread from which five or six tentacles project like the arms of an octopus. When stretched, the narrow body is 1 centimeter or more (½ inch) long.

The name hydra originated with the monstrous, nine-headed serpent of Greek mythology. Eight of its heads were mortal, but the ninth grew back as two heads each time it was severed. Hercules finally slew the monster by burning its heads.

Modern hydras are harmless but more strongly regenerative than their namesake. When any or all of their tentacles are cut off, they regenerate them. In this they resemble their indestructible associates, the flatworms (chapter 6).

When a small animal swims nearby, hydra grasps it in a tentacle embrace. The tentacles contain special cells, called *nematocysts*, which discharge poisonous barbs. Then the tentacles push the poisoned prey into the mouth to the digestive cavity. There it is digested and fed to cells lining the cavity. Conveniently, the body wall of hydra is only two cell layers thick. Every cell is within easy reach of the food.

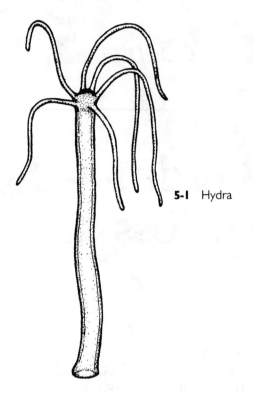

5-1   Hydra

When a hydra wants to move, it creeps slowly on its base. If it wants to move faster, it somersaults, first grabbing the substrate with its tentacles, then its base. Alternatively, it lassos objects with its tentacles, then draws its body towards them.

When disturbed, a hydra rapidly contracts into a small speck of lightly tinted jelly. It feels the slightest touch, responding with signals through a network of nerves. The nerves activate muscle cells in the trunk and tentacles, causing contraction.

Reproduction is sometimes sexual but often not. In asexual reproduction, one or two small hydras bud from the old body just as limbs bud from a tree. These break off in a few days to form new animals. In sexual reproduction, the sperm from a male testis joins the egg from a female ovary to generate a new animal. If reproductive organs are present, you will see them as small bumps on the body. Sometimes a single hydra has both male and female organs.

Materials

- Living hydras
- Hand lens or stereomicroscope
- Watch glass or small beaker
- Snail or slug meat or *Daphnia*

- Toothpick
- Acetic acid or vinegar

FINDING HYDRA

Look for hydras in lakes or slowly moving streams. They usually attach to water weeds a few inches below the surface, but they are hard to recognize. Movements cause them to withdraw, changing their shapes. Then you can feel and see them as faint gray or brown, jellylike masses. Break off some of the weed. Hold it quietly near the water surface, or place it in a bottle of lake water. If hydras are present, they will soon extend their bodies perpendicular to the weed surface. If you cannot find any, you can order them from a biological supply house (Appendix B).

EXAMINING AND FEEDING HYDRA

Place a live hydra in a watch glass or small beaker, and observe it with a hand lens or stereomicroscope. Watch for spontaneous movements of its tentacles. Now drop a tiny piece of snail or slug meat upon the tentacles. Alternatively, add living *Daphnia*, a small aquatic animal, to the water. Watch the tentacles of hydra slowly reach out to grasp the food. Then watch them move it into the mouth at the center of the tentacles, and from the mouth into the expanded gut for digestion. If not completely digested, a portion will be excreted back through the mouth.

OBSERVING THE DISCHARGE OF NEMATOCYSTS

You can see the discharge of slender, needlelike nematocysts by placing a hydra in water on a slide under a microscope. Allow the animal to extend. Stick a toothpick into acetic acid or vinegar, and touch this to the edge of the water. As acid diffuses about its tentacles, the hydra explodes numerous stinging barbs.

OTHER ACTIVITIES

Do your hydras have the regenerative ability of their Greek namesake? To find out, slice off one or more tentacles from each of several animals. Continue to feed the hydras with *Daphnia* and smaller organisms. What happens?

If your hydras lack ovaries and testes, you can purchase prepared slides of the organs or repeat some of the experiments described by W. F. Loomis in "The Sex Gas of Hydra," *Scientific American*, April 1959, page 145. The carbon dioxide gas found in stagnant water stimulates sexual reproduction.

Chapter **6**

Indestructible flatworms

Many of the less specialized forms of life regenerate missing parts. The planarian flatworms, for example, regenerate whole animals from each severed piece. If a head is cut off, the body grows a new head, and if the body is cut off, the head grows a new body.

More complex animals, such as crayfish, develop new limbs to replace lost ones, but cannot repair a severed body. Humans grow scar tissue, heal broken bones, and even rejoin cut nerves in favorable locations, but cannot replace severed fingers.

Biologists wonder why some animals regenerate parts of their bodies while others do not. Frogs and salamanders are often used for this research. Both are amphibians, but only the salamanders regenerate missing limbs as adults; and the severed limbs regrow only when their nerves are intact.

Though adult frogs fail to regrow lost limbs, tadpoles do regrow them. Probing the difference, researchers found that the nerve supply of tadpoles is greater than that of adults. With this in mind, they performed a surprising experiment. They opened an adult frog and moved the nerve of its hind leg to the site of a missing front leg, supplementing its nerve supply. What happened? This frog and others like it regenerated their front limbs, not perfect in size and shape, but they were there. Apparently the nerve secretes some chemical or chemicals that promote regrowth.

Returning to planarians, biologists who cut the animals' heads apart from their tails found that the head ends grew tails faster and better than the tail ends grew heads. Indeed, some tails refused to regenerate heads. What caused this predicament?

When the experimenters measured the oxygen consumption of each piece, they found that head pieces consumed more than tail pieces. Then they cut thin slices from just behind the eyes and far down the tail. Each upper slice

34

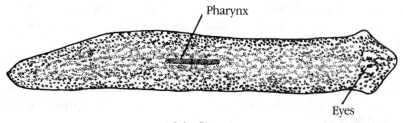

6-1   Planaria

grew two heads with no tail; each lower slice grew two tails with no head. Because the consumption of oxygen differs little on each side of a thin slice, the planarians seemingly could not tell the difference. Thus they failed to get organized.

Oxygen also helps organize the embryos of higher species. This can be shown by adding chemicals that alter their oxygen consumption. At critical stages, the chemicals cause many abnormalities of development.

Planarians reproduce either sexually or asexually (without sex). Asexual reproduction results from division, the head of the worm separating from the tail. The head then regenerates a new tail and the tail regenerates a new head.

Materials

- Living planarians
- Lake or stream water
- Hand lens or stereomicroscope
- Fresh liver or heart
- Razor blade
- Beaker or petri dish
- Dissecting needle or toothpick

COLLECTING, EXAMINING, AND FEEDING PLANARIANS

Look on the bottoms of rocks from streams or lakes for small, soft, jellylike blobs. These are the planarians. At first they are quiet, but soon a few will slip along, seeking darkness. Put the planarians in a jar of lake or stream water, and keep the jar in the dark. You can use tap water instead, if you first let it stand for several days in an open container.

Feed the planarians raw meat once or twice weekly. Watch them gorge themselves. When they finish eating, remove the meat and replace the stale water with fresh water.

When convenient, examine the planarians more closely. Each is 1 to 2 centimeters long (about ½ inch), is typically dark gray or brown but sometimes another color, and might have darker or lighter spots (FIG. 6-1). The two light spots on the triangular head are eyes, which appear crossed. Planarians obtain meat and animal matter through a *pharynx* that protrudes from the middle of the body. They move by secreting mucus over which they glide upon *cilia*. The cilia are microscopic, hairlike projections from their ventral (underside) surfaces.

If you watch the translucent planarians as they feed, you will see food accumulate in their digestive tracts. The *digestive tract* is Y-shaped, with one branch running forward and two backward. Numerous smaller tubes project from each side of the three main branches. Food enters through the pharynx at the middle of the Y.

Give the animals a bloody meal—fresh liver or heart—and watch the result. Examine with a hand lens or, if available, a low-power stereomicroscope. An intense red hue permeates the tract of all who eat. How many days does it take to disappear?

If you are careless in feeding the planarians, they will manage on their own. After using every speck of nourishment in the digestive tract, they digest the lining itself and then other cells of the body. They continue this for months, sometimes shrinking to ⅙ their former size.

BALANCE AND MOVEMENT TOWARD DARKNESS

Use a dissecting needle or toothpick to turn one of the planarians on its back. It flips over at once, demonstrating a balancing response found throughout the animal kingdom. On its own initiative, a planarian sometimes glides upside down across the upper surface of water.

Darken one side of your container for the planarians, and place a bright light on the other side. This procedure will confirm an observation made while collecting them: planarians retreat from light.

REGENERATION OF PLANARIA

To test regeneration, put a planarian on a smooth glass plate, and cut it in two or more pieces crosswise or lengthwise with a razor blade. Put each piece in a beaker or petri dish filled with lake or stream water. Cover and label the source of each piece. Keep the dish in a dark place. Examine the animal every two or three days, adding more water as the level goes down. The pieces soon start to regenerate; in about two weeks, each will have formed a smaller complete planarian.

Produce a two-headed planarian, if you wish, by slicing the head in two to the pharynx and keeping the two halves separated until regeneration is complete. Similarly produce two-tailed or many-tailed planarians by cutting their tails.

OTHER ACTIVITIES

Some biologists say they can transfer knowledge from one planarian (or other animal) to another. Generally, they train one animal to do a simple task, such as make a right turn in a T maze. Next they make an extract of its brain to inject into a second, untrained animal of the same species. Then they train the injected animal to do the same simple task. They find that the injected animal often learns faster than the original animal, suggesting that the extract of the brain helped.

You can try the same or a slightly different experiment on planarians. Purchase or collect both white and black species of fresh-water planarians, and keep them initially in separate containers. Train the black ones to do some simple task (this might not be easy). Then bring the two types together. I once mixed the two species and found, accidentally, that the whites were cannibals. They ate the other planarians. Perhaps if you bring the two species together, one will eat the other, including their small but well-trained brains. What effect will this have on learning the simple task? I am not sure, but the way to find out is to try the experiment.

Chapter 7

Collecting insects

"*K*aty did; Katy didn't." This call and the shrill chirp of crickets abound on summer evenings. Fireflies flash their lights; ants run through yards and kitchens; ladybugs eat aphids. Land, water, or air—it is hard to find a place without insects because there are more insect species than all other species combined.

Insects are *arthropods*, a word that means "jointed legs." Spiders, crayfish, crabs, centipedes, and millipedes are also arthropods, but differ from insects in the number of legs. There are six in insects, eight in spiders, ten in crayfish and crabs, and so many in centipedes and millipedes that we seldom count them. Arthropods also have *exoskeletons*, hard outer coats of armor that safeguard them. As they grow, they periodically shed, or molt, these skeletons.

Insect bodies have three divisions: *head*, *thorax*, and *abdomen*. The wings and three pairs of legs attach to their thorax. Most insects have four wings, some two, and others none. Most have *antennae* or feelers, two *compound eyes* that look like honeycombs, and tiny *spiracles* (holes in their sides) through which they breathe.

The class *Insecta* is divided into more than 30 orders (FIG. 7-1). Among these are the slope-winged insects called *Homoptera*, a group that includes cicadas, leafhoppers, and aphids. Cicadas are a favorite prey of birds and of a particular wasp called the cicada killer. Perhaps you have seen this large wasp sting a cicada and scoot up a tree for takeoff, carrying the cicada. It flies to its burrow, where it deposits an egg in the cicada. When the egg hatches, the larva eats its living but paralyzed host, devouring the vital organs last.

Cicada killers are members of *Hymenoptera*, the order of wasps, bees, and ants. Many of these are social animals, living together in hives and nests (chapter 9). Hymenopterans are the only insects with "stingers," a point of which most of us are aware.

38

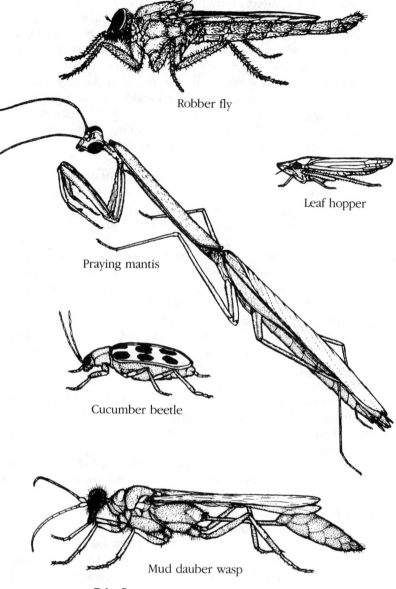

Robber fly

Leaf hopper

Praying mantis

Cucumber beetle

Mud dauber wasp

7-1   Representatives of five insect orders

Other social insects are in the order *Isoptera*—the termites. Termites eat wood but depend upon protozoans and bacteria in their intestines to digest it. Many species have separate castes of workers and soldiers. The soldiers have enormous heads which they bang against wood to warn others of danger. I once broke open some mud tunnels of termites along a tree trunk in the Virgin Islands. At each break, the big-headed soldiers gathered to defend the colony, running toward me when I knocked on the trunk.

For beauty among insects, most people choose the order *Lepidoptera*—the moths and butterflies. Butterflies fly during the day and have antennae that are slender and knobbed. Moths fly mostly at night and have antennae that are feathery or hairlike. Both suborders have caterpillars, and moths have cocoons.

A common pet in biology classrooms is the praying mantis, a member of the *Orthoptera*. This order includes other insects with chewing mouthparts: grasshoppers, roaches, katydids, crickets, and walking sticks. You can find or purchase the egg cases of the praying mantis (Appendix B). In the spring, one small mantis after another will emerge and begin eating its brothers and sisters or other insects. The young molt several times to reach an adult size of up to 10 centimeters (4 inches).

The largest group of insects is the hard-winged *Coleoptera*—an order of beetles and weevils. Some of these have distinctive, well-chosen names, such as the May beetle, which typically appears in May, and the burying beetle, which moves dead mice and other animals to its burrow. Another is the cigarette beetle, a gourmet of tobacco, and another yet the drugstore beetle, which eats pepper, cork, glue, and most anything else it can get its feelers on.

Diptera is an order of flies and mosquitoes, two-winged insects which most of us could do without. They are pesky and sometimes transmit diseases, such as malaria, dysentery, and tuberculosis. Their blood-sucking habits are especially distressful for livestock and wildlife.

The order *Ephemeroptera* contains mayflies. The young live as nymphs in water for one or two years, then come to the surface to molt and fly out in great mating swarms. The adults mate and die within a few hours or days, providing food for fish and bait for fishermen.

Lice are small, flattened insects that cling to the skin, where they resist being scratched off. They belong to the order *Anoplura*. Biting lice live mainly on birds and sucking lice wholly on mammals. The body and head louse of humans transmits typhus and relapsing fever.

Fleas are also bloodsuckers. They belong to the order *Siphonaptera*. Their bodies are flattened laterally (side to side), allowing them to run and leap quickly through hair. The most serious disease transmitted by fleas is plague, a sickness with fevers, chills, pains, and vomiting that killed a quarter of all Europeans in the fourteenth century. Rats, ground squirrels, and other rodents still carry fleas that sometimes transmit plague to humans.

The order *Odonata* contains dragonflies and damselflies, the nymphs of which are aquatic. Dragonflies are large, fast fliers at up to 100 kilometers per hour (60 mph). With their enormous compound eyes, they easily spot other insects, scoop them up with their legs, and often eat as they continue flying.

We have considered eleven orders of insects, omitting others. If you want to learn more, check your local bookstore or library.

Materials

- Insect net
- Small boxes or jars

Refrigerator with freezer, or carbon tetrachloride, medicine dropper, and
- Anesthetizing jar
- Insect pins
- Mounting board
- Collection box

PREPARATION OF EQUIPMENT

You need a net to catch flying insects, such as butterflies. Buy it, or make it from
an old broom handle, some wire (such as a coat hanger), and some mosquito
netting or very thin cloth. Bend the wire into a circle 25 centimeters (10 inches)
or larger, and attach it to the stick with nails or more wire. Sew the netting to the
wire, putting an extra thickness at the top to resist wear. Make the net nearly 1
meter (1 yard) long (FIG. 7-2).

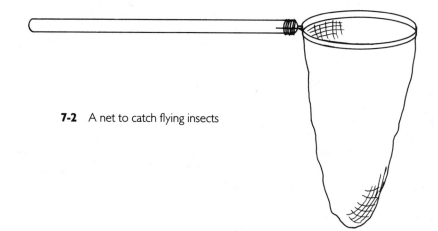

7-2   A net to catch flying insects

You also need a refrigerator with a freezer, or an anesthetizing jar, to kill
the insects for mounting. If you choose the anesthetizing jar, make it from a
wide-mouthed glass jar, such as a peanut butter or pickle jar. Mix a little water
with plaster of Paris, and pour 2 to 3 centimeters (1 inch) of this into the bottom
of the jar to dry. Alternatively, place a wad of cotton rather than plaster in the
bottom of the jar.

Last, you need a mounting board to spread the wings of large insects. Make
this from Styrofoam or soft wood as illustrated in FIG. 7-3. Use a longer board if
you want to mount several insects at one time. The slight, troughlike angle of
the mounting portion gives a natural flying attitude to the wings of your insects.
Stick pins through their bodies and into the Styrofoam, corrugated cardboard,
or balsa wood placed at the bottom of the mounting board.

CATCHING AND ANESTHETIZING INSECTS

You can find insects almost anywhere at any season, but the best time to collect
is late summer or early fall. Take the net, boxes or jars to hold insects, and an

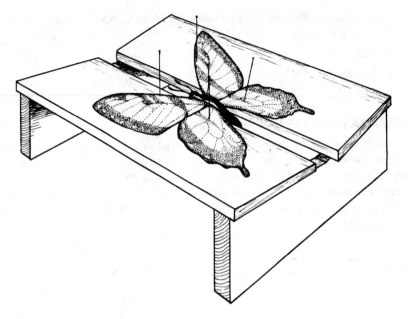

7-3   Mounting board for winged insects

anesthetizing jar (if desired), and go for a hike. Catch butterflies, bees, flies, and flying grasshoppers by sweeping them into the net. Flip the net quickly over the wire loop to imprison them. Pinch the thorax (the middle body division) of butterflies and moths. This weakens them, preventing them from flapping and damaging their wings when they are removed from the net.

If you plan later to freeze the insects, put them temporarily in collecting jars or boxes. Otherwise, put them in an anesthetizing jar now. For anesthetizing, add a few drops of carbon tetrachloride to the plaster or cotton in the jar, put a circular piece of cardboard over the plaster or cotton, drop the insects on the cardboard, and screw on the lid. The insects pass out quickly and die.

CAUTION: Avoid breathing the fumes of carbon tetrachloride. Use the anesthetic under adult supervision and only when you are outdoors.

Do not place delicate specimens such as butterflies in the same jar with beetles or other large, hard-bodied insects. Beetles kick scales off the wings of butterflies.

If you also collect soft-bodied larvae, drop them directly into 70 percent isopropyl or ethyl alcohol, and keep them in small vials.

MOUNTING AND LABELING INSECTS

Mount insects on the same day you freeze or anesthetize them; otherwise they dry in odd positions. Do not use regular straight pins for mounting. They are too thick and tend to rust. Purchase special, long, narrow insect pins from a supply house (Appendix B). Stick one of these through each insect, leaving enough pin above the specimen to grasp. Place pins in the middle of the thorax or slightly to the right in most specimens. Pin beetles close to the base of the right wing.

Mount butterflies, moths, and other winged specimens by pinning their wings open (FIG. 7-3). To ensure complete drying in this position, leave large specimens on the board for at least a week, and smaller specimens for a few days. With other insects, mount the legs in a natural position. Do this by sticking the pin with the insect into a small cardboard box until the feet touch the cardboard. Pin the legs in position, and allow the specimen to dry. Alternatively, pin the insect onto Styrofoam.

When the insects are dry, put them in a cigar box or glass-topped display case. Place a few mothballs in the box to keep out pests that might destroy the collection. These pests are fellow insects looking for dinner.

Find the order, common name, and scientific name of each insect. Arrange and label the insects by orders. By organizing the insects in categories and reading their life stories, you will learn much about this largest group of animal species.

OTHER ACTIVITIES

You can also learn about insects by conducting experiments. Here are three:

1. Refrigerate a few insects. Chilling reduces their metabolism, slowing or stopping their movements. Heat quickly revives them.

2. Place flies in a container filled with cigarette smoke. Smoke anesthetizes and might even kill them, as it sometimes (more slowly) kills humans.

3. Estimate the frequency of wingbeats by comparing the wing sounds of insects with the hum of tuning forks or piano cords. The forks and cords vibrate at known frequencies. When pitches of the insects and musical instruments are equal, so are the frequencies.

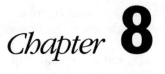

Chapter **8**

Emergence
of moths

Female moths attract males from great distances—sometimes up to ten kilometers. Biologists suspected for years that this attraction came from scent glands. In the open, females easily draw males, but in odorproof jars, they do not. Chemists isolated the moth perfume in the 1950s. To do so, Adolph Butanandt imported one million silkworm cocoons from Japan and Italy to his laboratories in Germany. He and his associates dissected scent glands from the females, then placed males near various fractions of the glands. When brought to the correct fractions, the males fluttered their wings in anticipation. With this knowledge, Butanandt eventually isolated 1.6 milligrams of the smell and found its simple formula: $C_{16}H_{30}O$.

Having attracted the male and been fertilized, the female distributes her *eggs* on a food source. Usually she fastens them to a plant with an adhesive (FIG. 8-1). In occasional species, she scatters them without attachment, sometimes as she flies.

The egg-sized *larvae*, or caterpillars, eat their way out of the eggs, then continue eating. Successively growing and molting, they attain the sizes most of us know. Molts occur each time the animals outgrow their skin. The larvae molt by first secreting a fluid that dissolves and separates the skin; then they expand their bodies to break free.

The full-grown larvae of most moths spin cocoons. The cerebral hemispheres of their small brains coordinate the intricate weaving. When experimenters remove one hemisphere of a silkworm, the worm spreads silk wherever it crawls. When they remove both hemispheres, the worm crawls about but makes no attempt to spin a cocoon.

Inside their cocoons, larvae become dormant and reorganized as *pupae*. They do not eat or visibly excrete. They mature during the winter, ready to break out in the spring as adult moths.

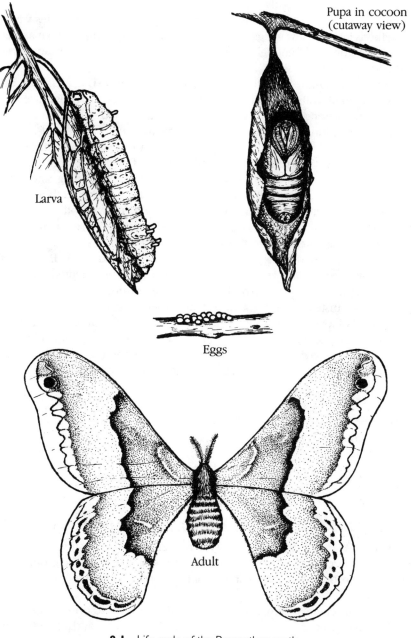

Pupa in cocoon
(cutaway view)

Larva

Eggs

Adult

8-1 Life cycle of the Promethea moth

The wings of a freshly emerged moth are crumpled and limp. While hanging from its cocoon, the moth sucks in air and contracts muscles which force fluids into the new wings. In about 30 minutes, the wings reach maximal size and become stiff enough for flight. Then it flutters its wings and, if so inclined, takes off.

Males often have large feathery *antennae* (feelers). These contain the organs of smell with which they locate females whose antennae are less feathery. The antennae also help biologists separate moths from butterflies. Butterflies have longer, thinner antennae with small knobs at the ends.

Both sexes of the moth have furry bodies, but the abdomen of the female is bigger. Her bigness results from a heavy load of eggs, not from a hearty appetite. Indeed, some adults have no mouthparts for feeding and must depend on food stored in their bodies as larvae. This food keeps them active long enough to mate and lay eggs. Other moths, such as hawk moths, have long sucking tubes through which to sip the nectar of flowers. When not in use, each moth coils its tube beneath its head.

What makes moths change from larvae to pupae to adults? V. B. Wigglesworth suggested the answer in the 1930s when he experimented, not with moths, but with a blood-sucking bug named *Rhodnius*. He decapitated the young bugs. Instead of dying, they transformed into little adults. Further experiments showed Wigglesworth that the removal of the *corpora allata*— two tiny glands projecting from each animal's brain—produces adults. The glands secrete a *juvenile hormone* that affects molts and metamorphosis, the change of one stage to another. This hormone was later found in many species of insects.

Materials

- Cocoons
- Box
- Scissors
- Tape

COLLECTING COCOONS AND WATCHING MOTHS EMERGE

Seek cocoons mainly in the fall or winter, after trees and bushes are bare of leaves. The cocoons hang from limbs and lie among leaves on the ground. Put them in a box outside your window, or if most of the winter has passed, bring them inside.

Carefully cut an oval window in one of the cocoons to observe the pupa inside (FIG. 8-1). Look for its future wings, mouthparts, eyes, and antennae; you can easily see their outlines. The pupa will later become a normal adult unless you damage it while cutting. Tape the oval "lid" back over the opening, and hang this cocoon with the others in such positions that the adults will have room to unfold their wings when they emerge. With continued attention and good luck, you may witness this warm-weather event, which sometimes comes early in the heat of a house.

Materials

- Caterpillars of moths
- Leaves to feed them

MOLTING AND SPINNING OF COCOONS

Watch for caterpillars in the late spring and early summer and adults in the late summer. Because they molt and spin cocoons the larvae are especially instructive. Collect a few. Feed them continually upon leaves of the kind of trees from which you obtained them. Observe them molting and spinning cocoons. Watch them deposit layer upon layer of silk.

OTHER ACTIVITIES

The brightly colored butterflies also deserve attention and so do their pupae called *chrysalides*. You may find the naked chrysalides under leaves or see them hanging from branches or fence posts. They look about like moth pupae from which the cocoons are removed, but usually they show more color. Collect the chrysalides and make follow-up studies of their life cycles.

When you have adult moths or butterflies, look closely at the scales on their wings. Wipe off a few of the tiny, powdery, overlapping scales with a paintbrush. Put these on a slide to observe under a microscope. As you examine different scales, you will see different colors and shapes for different species. Most of the colors come from wastes deposited in the developing pupae.

Chapter 9

Ant colonies

"Go to the ant," said Solomon, ". . . consider her ways, and be wise. . . . She prepares her food in summer, and gathers her sustenance in harvest."[1] Solomon was probably describing *harvest ants* that husk and store seeds underground, preparing for winter. They eat the seeds when other food is scarce.

Another group of ants, the *fungus growers*, does more than harvest. They plant, fertilize, weed, and protect gardens of fungus. The new queen starts a garden by carrying fungus to her nest. Her offspring cut, gather, and chew leaves to use as compost upon which the fungus continues to grow.

Several species of *dairying ants* collect aphids, mealybugs, or other small insects into herds for milking (FIG. 9-1). The "milk" is a secretion of honeydew obtained when ants stroke the backs of their "cows" with their antennae. The ants collect the eggs of aphids, and take them to their nests. Here they carefully tend them and, when the eggs hatch, place the offspring on suitable plants—often corn. The aphids partially digest the sap of corn and pass a large part of it to the ants.

In deserts where the supply of honeydew is seasonal, *honey ants* store it in their own bodies. Workers of these ants regurgitate the sweet liquid into the mouths of a few. In these, the abdomens swell until they are spherical and thin. Unable to move, the swollen ants hang from the ceilings of specially constructed chambers. In time of need, they discharge sweets to other ants.

In the tropics, pursuing birds call attention to a moving *army* of *ants* several meters wide and sometimes hundreds of meters long, blackening the ground. The density of ants at the center forces many to the sides, where they swarm ahead. The swarm flushes prey, such as insects, into the main body of ants, where the prey is torn apart and carried to the rear. Few escape. Even horses left tied are eaten.

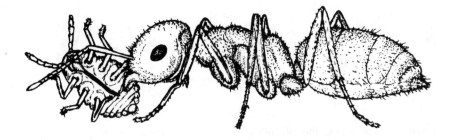

9-1   Worker ant carrying aphid

Army ants seem both wonderfully intelligent and amazingly stupid. They march in an instinctively disciplined manner, rescue fellow ants in distress, make living bridges across difficult terrain, and generally show a high degree of social order. Yet their unalterable marching behavior can be their undoing. If once started in a circular path on flat ground, each ant follows the one immediately in front until all perish from lack of water.

Materials

- Trowel or shovel
- Ants in your yard

OBSERVING THE LIFE CYCLE AND NEST BUILDING

On a warm, humid, summer afternoon you might happen upon the craziest mess of ants you have ever seen. In your yard and those of neighbors, thousands of ants run about in total confusion. This is the time of *swarming* and mating. Winged ants as well as workers are present. The large-winged ants are new *queens*; the countless, smaller winged ants are *males*. Continue to watch the disorderly swarm. After a few hours, the queens and their suitors fly and mate in the air.

Male ants perish a few days after mating, but the fertile queen lands, disposes of her wings, and tunnels into the ground (FIG. 9-2). Sealed from the outer world, she waits alone till her eggs mature. The decomposition of flight muscles provides her only nourishment.

9-2   Wingless queen

Ant *eggs* pass though *larval* and *pupal* stages before *adulthood* (FIG. 9-3). The queen feeds saliva to the first batch of wriggling larvae, which might or might not spin cocoons, depending on the species. Then all become quiescent pupae. Later the pupae transform into workers that dig to the surface to find food.

9-3   Egg, larva, pupa, and adult

The queen's sole duty becomes that of laying eggs during the dozen or so remaining years of her life. Workers nurse the young—placing them in areas of proper warmth and moisture, licking larval secretions, and helping new workers emerge from cocoons. Other workers obtain plants or insects upon which the colony feeds, regurgitating the partially digested food from ant to ant.

To observe some of these events, dig into an ant nest. If you reach the larvae and pupae, identify them. Worker ants pick them up to move them to safer locations. Watch during the next hours and days as the ants rebuild their nest.

Watch also as the workers collect food. A trail of ants moves outward from the nest to the food while others return with food by the same trail. As the ants move, they touch their abdomens to the ground to deposit a *pheromone*. Pheromones are species-specific odors that convey messages, such as "danger" or "come mate with me" or "this is the trail." If you want to cause pheromonal confusion, rub your finger or shoe across the trail of ants. This will remove the odor that guides them. Watch what happens as the ants come to the break in their trail.

Materials

- Ant's nest containing one or more queens, workers, and whatever else you find
- Observation nest that you buy or make from wood and glass panes
- Trowel or shovel
- Collecting jars

STUDY OF ANTS IN AN OBSERVATION NEST

Select an ant nest—such as that of the common black or red ants in your yard—and dig deep with a trowel or shovel around its main opening and to each side. Spread the dirt on the ground, or better, on an old bed sheet, seeking first a queen or queens. These are elusive, but with effort you should find one. Place workers, winged males, and young, or whatever is available, in one jar, and the large queen or queens in another. Both jars should contain dirt to be used in filling an observation nest and to prevent drying of the ants.

Upon return to the laboratory, place the ants with earth in an observation nest, such as those available from biological supply houses and toy stores

(Appendix B). If you prefer, make your own nest from two panes of glass, each about 20 by 30 centimeters (8 by 12 inches), and a tight-fitting wooden frame (FIG. 9-4). Dampen the soil of the chamber at weekly intervals, but do not make it so wet that mold grows. Feed the ants with honey and an occasional soft-bellied insect or table scraps. Give only small amounts of food.

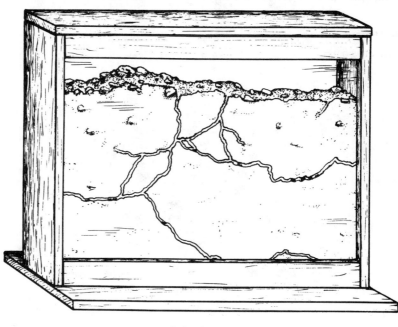

9-4 Ant nest

Without disturbance, observe the ants as they construct their nest. Notice how they use their legs and mouthparts in excavation. Notice the reaction of workers to the queen. Introduce one or two strange ants to see the distress this causes. What weapons are used in fighting? If not killed, the new ants acquire the nest odor in a few hours and will then be accepted as fellow workers. Watch for the regurgitation of food from ant to ant. This allows a transfer of nutrients and perhaps of pheromones that regulate the size and composition of the colony.

OTHER ACTIVITIES

Termites are another of the social insects worth studying. In the spring or summer, examine decaying tree stumps and logs to find a colony of the white-bodied termites. Look for the different castes—soldiers and workers being the most common—then cut into the log to study the tunnels.

Put the termites in a jar with some of their wood. Upon returning to your home, make some saline solution by dissolving ½ teaspoon of salt in 1 cup of water. Cut off the abdomen of one termite, taking care that none of the other termites escape. Push out the intestine, and tease it apart in a drop of the saline solution on a slide. Under a microscope, look for the protozoa and bacteria that allow termites to digest solid wood. The termites themselves cannot directly digest the cellulose of wood. Instead, the protozoa and bacteria digest it, passing some of the digested products to their hosts.

Endnotes

1. Proverbs 6:6,8

Part 3

Animals with backbones

Chapter 10

Fish, aquaria, and snorkeling

"*M*ay 17. Norwegian Independence Day. . . . I am cook today and found seven flying fish on deck, one squid on the cabin roof, and one unknown fish in Torstein's sleeping bag." Thus reads the log of Thor Heyerdahl, leader of the Kon-Tiki expedition, whose group was rafting across the Pacific.[1]

The explorers lured flying fish at night by lighting a paraffin lamp. The fish shot above the raft, striking the bamboo cabin or sail, tumbling to the deck. "Sometimes," said Heyerdahl, "we heard an outburst of strong language from a man on deck when a cold flying fish came unexpectedly, at a good speed, slap into his face. . . . But the unprovoked attack was quickly forgiven . . . for, with all its drawbacks, we were in a maritime land of enchantment where delicious fish dishes came hurling through the air."[2]

Most of us see more fish in supermarkets than in oceans. We eat herring, tuna, salmon, or codfish in place of flying fish. The tasty herring, for example, are Atlantic fish, abundant in the North Sea and north of Cape Cod, and packed as sardines. Pacific canneries use pilchards, a relative of herring. In freshwater, the desirable food and game fish are trout, bass, pike, sunfish, perch, and catfish.

All fish are divided into two classes (FIG. 10-1):

1. The *cartilaginous fish* (Chondrichthyes) are those with skeletons of cartilage—mostly sharks and rays. You may feel similar cartilage in your own body by bending your ear or touching your larynx (Adam's apple). Because we see sharks often on television, I shall describe only the bottom-dwelling rays with gigantic fins that constitute most of their body. They swim by slowly flapping these fins like wings. Their two eyes are on top of the head, and the mouth and gills on the bottom.

2. The *bony fish* (Osteichthyes) are all those previously mentioned except

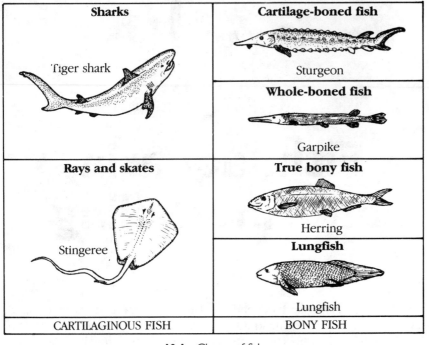

Sharks	Cartilage-boned fish
Tiger shark	Sturgeon
	Whole-boned fish
	Garpike
Rays and skates	**True bony fish**
Stingeree	Herring
	Lungfish
	Lungfish
CARTILAGINOUS FISH	BONY FISH

10-1   Classes of fish

the sharks and rays. Their bony skeletons are hard like ours. Their gills have a single exterior opening on each side of the neck instead of multiple openings, as in sharks and rays.

Materials

- Fish
- Dissecting tray
- Dissecting tools
- Hand lens or microscope (optional)

ANATOMY OF A FISH

Next time you go fishing, do more than eat your catch. Take time to examine one of the fish, beginning with its surface (FIG. 10-2). Look for two sets of paired fins: *pectoral fins* behind the gills and *ventral fins* on the underside. Fossil evidence suggests that the fins of fishes evolved into the legs of amphibians about 300 million years ago. Had there been more than two pairs of fins, higher animals might now have more arms and legs. Look also for *spiny dorsal* and *soft dorsal fins* on the back and an *anal fin* on the underside. The soft dorsal and anal fins are near the tail, which has itself developed a *tail fin*, important for steering and locomotion.

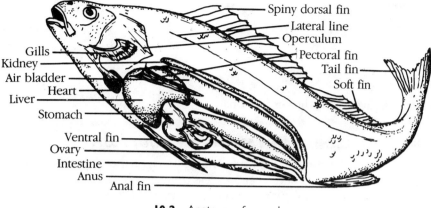

Gills
Kidney
Air bladder
Liver
Heart
Stomach
Ventral fin
Ovary
Intestine
Anus
Anal fin

Spiny dorsal fin
Lateral line
Operculum
Pectoral fin
Tail fin
Soft fin

10-2   Anatomy of a perch

Find the *operculum*, the hard plate that covers the *gills*. Lift the operculum and flush water over the gills to see their feathery structure. They provide a large, blood-filled surface through which oxygen enters from water, and carbon dioxide exits from blood.

Notice the *lateral line*, a light band running along both sides of the fish from the operculum to the tail. This is a sense organ for detecting movements of water. With it, fish spot and avoid predators and other objects in the water.

Most fish have *scales* of different shapes that help ichthyologists identify their species. These scales often show annual rings of growth similar to those in trees. They grow faster in the summer. If you have a hand lens or microscope, examine the scales more closely. How old is your fish?

Open the fish. Find its *heart* at the head end, beneath the gills. The heart has two chambers, an *atrium* and a *ventricle*. Are the chambers still beating? They may be, if you opened the fish when it was first killed. The heart pumps blood from the atrium to the ventricle to the gills, then through arteries to the rest of the body and back through veins to the atrium. In contrast to fish, amphibians and most reptiles have a three-chambered heart (two atria and one ventricle) and birds and mammals have a four-chambered heart (two atria and two ventricles).

Follow the short, tubular *esophagus* from the mouth into the *stomach*. Continue to the *intestine*, where food is digested and absorbed, and to the *anus*, where wastes are excreted. The anus is near the front edge of the anal fin. Look also for a *urogenital* opening between the anus and fin; this connects to the two *ovaries* of the female or the two *testes* of the male, and to a pair of long, dark *kidneys*.

Find the swim or *air bladder* near the kidneys and backbone. This elongated, air-filled sac helps the fish suspend itself in water, floating without effort. To ascend, the fish fills the bladder with gases from the blood. Lungfish gulp air into the bladder, using it as a primitive lung.

Materials

- Aquarium
- Gravel or sand
- Fish
- Plants
- An aerator and heater (optional)

PREPARING AND MAINTAINING AN AQUARIUM

Over 20,000,000 people in the United States have aquariums, and with good reason. Where can you find quieter or less troublesome pets than fish? Their beauty adorns many city apartments in which other creatures would fail or be thrown out by the manager.

Large aquaria of 5 to 20 gallons have stable temperatures, suiting them for a variety of fish (FIG. 10-3). Settle for less if your pocketbook forbids extravagance.

Shiner

Cabomba

Darter

Sagittaria

10-3 An aquarium

Thoroughly wash enough gravel or sand to cover the bottom of your tank to a depth of 2 or 3 centimeters (about 1 inch). Then add water until the tank is nearly filled. If convenient, use pond, lake, or stream water. Otherwise, use tap water that you have kept in an open container for three or four days.

Add living plants to provide shelter, some food, attachment for eggs, and beauty. I suggest *Vallisneria, Sagittaria, Myriophyllum, Cabomba,* or *Anacharis,* or some combination of them. Aquarium-grown plants adapt and endure better than pond plants.

Locate the aquarium in strong, diffuse light. To prevent an excessive growth of algae in the water, keep the aquarium out of direct sunlight. After the water has cleared and the plants have started to grow, add fish and a few snails.

The snails are scavengers that remove uneaten food. Also add a cover to the tank to keep the fish from jumping out.

If you want native fish, catch small ones with a seine. For color, try to find red-bellied daces. Then add shiners, minnows, darters, killifish, or others. Some may be predators, so don't be surprised if the smaller fish disappear in a few weeks. Visit your local bookstore or library to find pictures and descriptions of the fish and their appetites.

You may prefer tropical fish, which are generally more colorful. If you are a beginner, buy guppies. They are inexpensive and hardy. Then add platies, angelfish, gouramis, or others your dealer may suggest.

Never overcrowd fish nor put them in a container with a small surface of exposed water. Under these conditions, carbon dioxide may accumulate to stifling proportions. In the usual rectangular tanks, use only one to two inches of fish per gallon of water. If the fish are somewhat crowded or if they normally live in aerated water, use an aerator, and if they are tropical, use an aquarium heater.

Food requirements vary with different species. For most fish, use one of the well-balanced, prepared foods sold at pet stores. Do not overfeed. If desired, supplement the prepared food occasionally with living insects or *Daphnia*, a tiny crustacean.

Materials

- Face mask
- Snorkel
- Bucket
- Hand lens

SNORKELING AMONG FISH

CAUTION: Swim in safe, clear water with a partner. Use adult supervision.

With a partner, swim or wade among fish and other animals to make direct observations. Use a glass-covered mask for the best vision, and a snorkel tube to allow you to keep your face immersed (FIG. 10-4). When you are ready to go snorkeling, spit on the inside glass surface of the mask. Rub the saliva evenly across the surface, and rinse it out with a little water. The remaining thin layer of saliva prevents the glass from fogging. Then fasten the mask snugly over your eyes and nose, and fit the mouthpiece of the snorkel between your lips and gums. Breathe through your mouth and the snorkel. While swimming with your face underwater, the snorkel tube projects above the water, allowing you to inhale air.

If you are in an ocean or large lake, waves occasionally wash over the air vent of the snorkel, causing water to pour into the U-shaped part of the tube below your mouth. This happens infrequently because your body floats on the

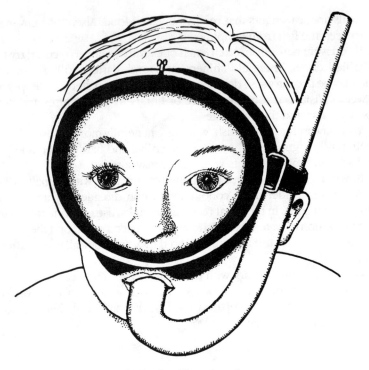

10-4 Snorkle and mask

surface of the waves. When it happens, abruptly exhale a big breath of air, blowing the water out the tube. Alternatively, choose a snorkel that has a ball in a cage at the opening of the tube. The ball acts as a floating valve, sealing the tube when water starts to enter it. With this type of snorkel, you seldom need to expel water.

If you are wading or swimming in a bay or ocean, collect small fish, sea anemones, starfish, or other animals for temporary study on shore. Use a hand lens to examine them. Watch how they move and feed. When you finish, put them back where you found them.

OTHER ACTIVITIES

You may fill your aquarium with creatures other than fish. Seine a pond or lake. You will catch water insects of many kinds, perhaps including the giant water bug, the male of which carries the female's eggs on his back. You may also catch tadpoles and aquatic salamanders. Each animal will provide hours of enjoyable looking.

Marine animals are easy to collect but more difficult to maintain than freshwater animals. If you want to try a marine aquarium, place only one small organism per gallon of seawater in a glass, rustproof tank. Add a few shells for hiding places and, if the aquarium is large, one or two plants. If you live near the ocean, replace the stale seawater daily with fresh seawater. If you live inland, purchase a marine salt mixture to combine with distilled or deionized water as directed on the package. Feed your animals in a separate container to avoid contamination of the aquarium by uneaten food. As water evaporates from the aquarium, add enough distilled or deionized water to restore the volume lost.

Endnotes

1. Heyerdahl, Thor. *Kon-Tiki: Across the Pacific by Raft*, p.13. Chicago: Rand McNally & Company, 1984.

2. Heyerdahl, p. 114.

Chapter 11

Terraria for amphibians and reptiles

T here is a fish—the sea robin in the Gulf of Mexico—with fins so strong that it crawls upon mud, seeking crabs, worms, and mollusks. But it breathes by gills, so it cannot survive on land.

There is another fish—the lungfish in Africa, South America, and Australia—with an air bladder developed as a working lung. The lung keeps the fish alive even when the water of its pools evaporates, leaving mud. But the fish swims mainly with its tail, having no legs with which to crawl on land.

If the lungfish had legs or the legged fish had lungs, the resulting animal could move across land from pond to pond. Are there creatures today with both features? Yes, these are the *amphibians*: salamanders, frogs, and toads.

What are amphibians? They are animals with double lives. As larvae they swim in water, breathing with gills, but when the larvae mature, most *metamorphose*. They lose their gills and gain lungs and limbs, enabling them to live on land. One example of metamorphosis is the change of tadpoles into frogs.

Amphibians, such as salamanders, resemble reptiles, such as lizards. But reptiles are permanent residents of land. They have no larval stage and no gills. Instead they have lungs, waterproof skin, and shelled eggs. Having lungs, they breathe air rather than water; having a dry, waterproof skin, they evaporate little water, conserving it for use by their bodies; and having shelled eggs, they can lay them far from water. Some snakes and lizards have gone a protective step further by giving birth to live young.

In the days of the dinosaurs, reptiles were the most important of all land animals. Now there are fewer species and smaller numbers. Today's representatives are crocodiles, alligators, turtles, tortoises, lizards, and snakes.

Reptiles and amphibians are often described as "cold-blooded" animals, but they are not always cold. Their body temperatures vary with those of their surroundings. They warm considerably, for example, while sunning on rocks. They seek sunshine when they are cool and shade when they are warm, regulating their temperatures for optimal activity.

Materials

- Net with a long handle for adult amphibians and lizards
- Seine for tadpoles and aquatic salamanders
- Noose for snakes
- Jars or muslin bags

COLLECTING AMPHIBIANS AND REPTILES

Amphibians and reptiles are shy. They conceal themselves under rocks, bark, and leaf litter, and hide in rotting logs. To find them, take a walk in the woods, stopping to turn over rocks and logs as you go.

To catch amphibians and lizards, use your hands or a stout net of 3-millimeter (⅛-inch) mesh cloth with a handle the length of your leg. Seek amphibians along the shores of swamps or lakes or in other moist surroundings. More come out after a warm rain. Locate frogs and toads by their calls at night. If you flash a light in their eyes, they often remain stationary till caught. Seine long-established ponds for tadpoles and aquatic salamanders.

To catch snakes, use a leather noose bound to a wooden handle (FIG. 11-1). Alternatively, pin them down with a foot or forked stick, and to avoid snakebites, grab each just back of the head.

CAUTION: Use adult supervision. Also, be familiar with and avoid poisonous species. Leave these for professional herpetologists.

11-1 Snake noose

Put living specimens in jars or muslin bags till they can be brought to better housing. Place dampened moss or leaf litter in these containers.

Materials

- One or more aquarium tanks, battery jars or other observation chambers

PREPARING TERRARIA

A terrarium is a cage or tank in which live amphibians and reptiles are kept for observation. Requirements of individual animals vary. Most come from either a swamp, woodland, or desert habitat and will feel at home in similar environments.

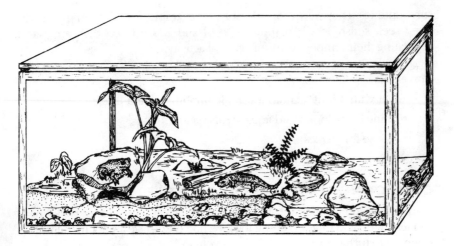

11-2   Swamp terrarium for amphibians

Permanent terraria should be large. A container of 5 to 20 gallons is ideal. Use an aquarium tank, battery jar, or other transparent chamber (FIG. 11-2).

For amphibians and wetland reptiles, cover the terrarium with a glass plate, propped up slightly to allow entry of air. This prevents animals from escaping and maintains a higher humidity. For desert animals and most snakes, cover it instead with a tight-fitting, screen-wire lid.

Use both native animals and plants in terraria but do not overcrowd either. Include areas of seclusion, such as rock overhangs or small hollow logs.

SWAMP TERRARIUM

Many amphibians and a few reptiles do well in watery surroundings. Pour 5 to 10 centimeters (2 to 4 inches) of water into an aquarium tank. Fill one end above the water line with rocks, sand, and a little soil. Add both land and water plants.

For amphibians in your swamp aquarium, choose tadpoles, frogs, or salamanders. For reptiles, choose turtles or water snakes.

WOODLAND TERRARIUM

Add 2 to 3 centimeters (about 1 inch) each of gravel, sand, and soil, in that order, to your tank. Then pour in water to the base of the soil. Add woodland plants—such as mosses, ferns, and liverworts—and sink a pan of water into the soil. Most animals will drink from this pan, but some lizards will drink only from sprinkled plants. Snakes prefer a water dish that permits them completely to submerge. When the terrarium is ready, put toads, tree frogs, woodland salamanders, lizards, or snakes in it.

DESERT TERRARIUM

Fill the bottom of the tank with 5 to 8 centimeters (2 to 3 inches) of sand and place on it two or three large rocks. Add cactus and a shallow pan of drinking water. Keep the lower level of sand moistened with a small amount of water.

Many desert animals—such as those from the southwestern United States—need warmth. You can supply this by suspending a light bulb over one end of the cage. Keep the heated end of the cage at about 30°C (87°F). When the terrarium is ready, put desert lizards or snakes in it.

FOOD FOR YOUR ANIMALS

During warm weather, most amphibians and reptiles feed daily, if there is plenty of food. In captivity, two or three feedings weekly will suffice, and one weekly for snakes. Amphibians and reptiles use less energy than birds and mammals, especially in cool weather, so they eat less food. At winter temperatures, most eat nothing and survive.

What should you feed them? Mostly living insects, but this varies as indicated in TABLE 11-1. For more details, check your library.

Table 11-1   Food Preferences of Amphibians and Reptiles

Animals	*Preferred food*
Aquatic salamanders	Crayfish, fish, insects, and plants
Terrestrial salamanders	Insects and spiders
Lizards	Insects
Snakes	Some eat rodents, some eat earthworms, and some eat reptiles, fish, and frogs
Turtles	Some eat plants, some animals
Tortoises	Usually plants
Alligators	Crayfish, fish, and insects
Crocodiles	Meat

OTHER ACTIVITIES

Because amphibians and reptiles have little economic importance and lead secluded lives, they are less studied than most animals. Amateurs, therefore, have opportunities for worthwhile, sometimes original observations. Get a field guide to amphibians and reptiles, then watch them in the field and in terraria. Keep records of their habits, foods, and active periods.

Chapter 12

Observing birds

*H*umans have tried to imitate birds in self-powered flight with little success. Birds succeed because they have the *power* to move their wings, and *lightweight bodies* for the wings to lift. The power is provided by large breast muscles.

Despite the large muscles, birds are much lighter than other animals. You might have noticed, for example, how ducks float high in the water compared to you or other swimming mammals. They float high because their bodies consist mostly of air trapped in feathers, bones, and large air sacs within their bodies. Both the feathers and bones have hollow, air-filled shafts. You know how light feathers are, and you might know that the down feathers of birds are used to fill sleeping bags. As for the bones, those of a frigate bird with a 7-foot wingspan weighed only 118 grams (4 ounces); they were lighter than its feathers!

The *lungs* of birds connect to large *air sacs* that sometimes extend into the bones. Air passes through a *trachea* or windpipe to the lungs and posterior air sacs, then back through the lungs, anterior air sacs, and trachea to the outside. Oxygen mainly transfers to the blood during the return of air through the lungs.

Birds have no urinary bladder in which to store urine, no bladder in which to store weight. Instead, both the *kidneys* and *intestine* shift wastes into a common excretory passage—the *cloaca*. It ejects a white paste.

In addition, the embryos of birds develop in *eggs* laid outside the body, not in uteri. Also, there is only one *ovary* in most birds, not two as in mammals, and the reproductive organs are shrunken except in the reproductive season.

The heads of birds are small and lightweight in comparison with their bodies. They have no teeth or heavy jaws, only a bill. They grind food in a *gizzard* near the center of the body, not with teeth in the head. Thus they avoid the need for heavy tails to counterbalance their heads during flight.

Because they can fly, birds can readily improve upon or stabilize their environments by moving to sites where food, water, and climate are optimal. Most birds migrate.

Materials

- Binoculars
- Field guide to birds

OBSERVING AND IDENTIFYING BIRDS

When and where should you look and listen for birds? Usually the best season is spring and the best time is morning. The bright-colored males are singing, defending their territories, and courting the females. You will see different species in different locations, but the best places usually are in parks, at the edge of woods, and in marshes. In marshes, the best times are during droughts when birds congregate at water holes.

Binoculars help. If you buy binoculars, get those that you can hold steady, that provide adequate magnification, that admit sufficient light, and that allow you to scan a large territory. Bird watchers often choose binoculars with these approximate specifications: 7 x 35 with a field of 400 feet at 1000 yards. Seven is the magnification and 35 the diameter of the objective lens in millimeters. The field of view allows you to see an image that is 400 feet wide at a distance of 1000 yards.

On walks to and from school or work, around lakes, or on hikes through woods, identify as many birds as possible. Start with robins, starlings, sparrows, and pigeons. Advance to flickers, warblers, herons, and hawks. Note differences in appearance, referring to FIG. 12-1 and a field guide for marks of identification. Then observe variations of songs and behavior. Without disturbing the parents or young, locate nests with eggs and young birds (FIG. 12-2). By observing these from a distance with binoculars, you can begin your study of life cycles. There are several excellent field guides, including Roger Tory Peterson's *A Field Guide to the Birds* and *A Field Guide to Western Birds*.

Materials

- Sunflower seeds or other bird food
- Food container
- Bird bath

FEEDING AND WATERING BIRDS

Not all birds migrate in the winter to warm climates. Some stay behind where snow and ice make feeding difficult. At such times, you can help the birds and observe their behavior by feeding them.

Different species prefer different foods. In comparison feedings, however, most of the seed-eating birds choose mainly sunflower seeds, so you will

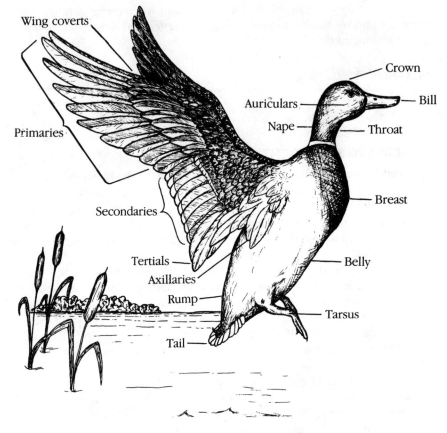

Wing coverts

Primaries

Secondaries

Tertials

Axillaries

Rump

Tail

Crown

Bill

Auriculars

Nape

Throat

Breast

Belly

Tarsus

I2-I   Identification marks of a bird

probably want to provide these seeds or mixtures that contain them. Other birds, accustomed to insects and worms rather than seeds, prefer suet—that is, beef fat or tallow obtained mainly around the kidneys. For an experiment, place several trays of different seeds and one of suet in the same location in your yard. Find which birds come to which foods and how often.

Spread the food on the ground or, better, on an elevated platform with a roof to keep off the rain and snow (FIG. 12-3). Place the platform near your window to observe it, but far enough from bushes and trees that predators such as cats cannot readily spring on the birds. If squirrels or other animals are getting the food, put the food container on a pole inserted through a large, downward-facing funnel. Place the funnel high enough to prevent the animals from leaping above it.

You can also attract birds by providing a place for them to drink and bathe. The container can be as simple as a pie pan or ceramic saucer refilled daily with fresh water and placed on a tree stump. For bathing, most birds prefer wider containers that are still shallow enough for them to hop about and dip their heads and wings in the water. The best containers are at least 60 centimeters (2

12-2 Five-hour-old killdeer

5 to 7 centimeters (2 to 3 inches). This is the usual size for concrete birdbaths. Keep the bath near enough to trees and bushes that birds can readily fly to it but far enough for them to see and avoid predators.

Materials

- Wing feather
- Down feather
- Microscope or hand lens

ANATOMY OF FEATHERS

Find and examine a wing feather (FIG. 12-4). Cut into the cylindrical *quill* at the base of the feather to see that it is hollow, reducing its weight. Then examine the *barbs* that project to each side of the quill farther up. Each barb runs parallel to the barbs on each side of it. Use a hand lens or, better, a microscope to see the finer structure of the feather. The barbs each have branches called *barbules*. Some of the barbules have *hooks* that fit over other barbules. The hooks hold the barbs and barbules firmly together, allowing the wing feathers to push as a unit against air during flight. Break the hold between two of the barbs, then pull them together as birds pull separated barbs together with their bills.

Look also at a down feather to see that it has no hooks. Down feathers provide insulation.

Materials

- Whole chicken
- Sharp knife
- Forceps or forks
- Scissors
- Cake pan or similar container

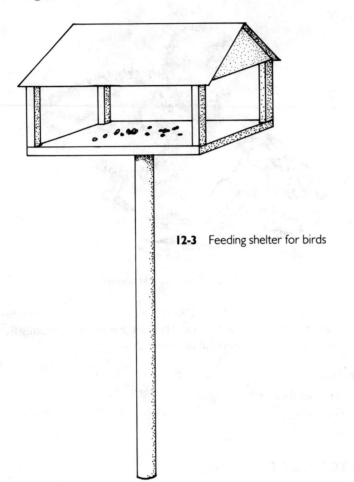

12-3   Feeding shelter for birds

ANATOMY OF FLIGHT MUSCLES

Obtain a whole chicken from a grocery.

CAUTION: If you plan later to cook and eat the chicken, you should complete your dissection in 15 to 20 minutes and wash your hands afterward. Chickens have bacteria that flourish when the birds are left too long in a warm room.

Place the bird in a cake pan or similar container, and remove the skin from its breast to expose two layers of flight muscles:

1. The *pectoralis muscle* is outermost and thicker. It originates on the keel of the *sternum* (breastbone) and inserts on the underside of the *humerus* bone in the wing. As you see from its attachments, the

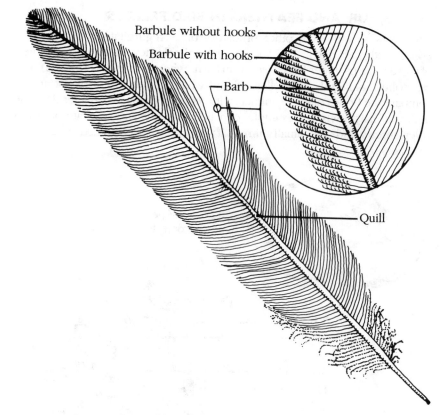

Barbule without hooks

Barbule with hooks

Barb

Quill

12-4   Anatomy of feathers

pectoralis pulls the wing downward during flight, lifting the bird. Pull the muscle and its tendon to show this effect.

2. The *supracoracoideus muscle* is under the pectoralis, next to the sternum. Follow the muscle and its tendon from the sternum to the back side of the humerus. As you see, the supracoracoideus pulls the wing upward during flight. Pull the muscle and its tendon to show this effect.

By having both of these heavy muscles on the underside of their bodies, birds gain stability during flight.

Cut into several bones, including the humerus. Notice the hollow structure that makes them lightweight. In some of the larger birds, such as vultures, the bones are internally reinforced by struts similar to those in aircraft.

Materials

- Bird pellets
- Forceps or teasing needle

BONES, FUR, AND FEATHERS IN BIRD PELLETS

Owls, hawks, and some other predatory birds regurgitate the undigested bones, teeth, hair, and feathers of the small animals they eat. They drop pellets of these remains under large trees or in barns where they roost (FIG. 12-5). You can collect and study the pellets to find what animals the birds eat and to compare the bones of these animals with those of humans (Chapter 24). The pellets of barn owls, for example, contain the skulls and separate bones of voles, shrews, and other small animals. If you prefer, buy sterilized pellets from suppliers (Appendix B).

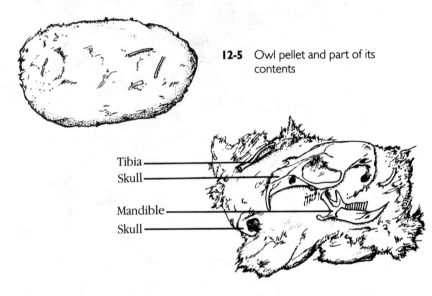

12-5   Owl pellet and part of its contents

Tibia

Skull

Mandible

Skull

By carefully sorting through the entire contents of a pellet, you will find virtually all the bones of one or more small mammals or birds. Identify these bones, arrange them in anatomical sequence, and glue them to a white card for display. Alternatively glue them in the shape of the standing animal.

OTHER ACTIVITIES

If you like to watch and talk about birds and to take field trips, you might want to join a group of experienced bird watchers. If so, telephone your local bird club, usually listed in the business section of the directory under "Audubon," or write to:

National Audubon Society
645 Pennsylvania Ave., S.E.
Washington, DC 20009

For an anatomical study through binoculars, examine the bills and feet of different birds. Notice the adaptations of each. Feet, for example, can be webbed for swimming as in ducks, elongated for wading as in herons, taloned for grasping prey as in eagles, curved for grasping branches as in sparrows, or feathered for cold climates as in ptarmigans.

Chapter 13

On the trail of mammals

*H*air and mammary glands—these distinguish mammals from scaly reptiles and feathered birds. Mammals have hair during some phase of life, and mammals nurse at their mothers' breasts.

There are three divisions of the milk-giving class *Mammalia: the* monotremes, marsupials, *and* placentals.

1. The *monotremes* of today are duckbilled platypuses and spiny anteaters. They lay eggs and show other similarities to reptiles; yet the young, when hatched, lick milk from mammary glands.

2. The *marsupials* are kangaroos, opossums, and other animals with pouches. Born small, the young scurry to the warmth of the mother's pouch for further development.

3. The *placentals* are horses, dogs, bats, whales, humans, and other animals that develop in a uterus inside the mother. The unborn attach there to a placenta, a vessel-filled structure through which oxygen and nutrients diffuse from the mother to the fetuses, and wastes diffuse from the fetuses to the mother.

Mammals succeed, as do other animals, by adapting to their environments, thus surviving and reproducing. Among the adaptations are camouflage and deceit. How many predators see a spotted fawn lying in a thicket dappled with sunlight? How many circumvent an opossum that pretends death when overwhelmed by danger? Biologists believe that the pretense is not a conscious act but a form of shock caused by fear—a faint instead of a feint.

Bats, woodchucks, and ground squirrels adapt to winter scarcities of food by hibernating. As winter approaches, they retire and become dormant, dropping their body temperatures to those of their surroundings, sometimes near the freezing point of water. They eat no food, have only a few heartbeats per minute, and breathe slower yet—sometimes less than once per minute. Bears, skunks, and raccoons sleep through the winter but show less change in temperature, heartbeats, and breathing.

Whales, porpoises, and seals adapt by swimming and diving. They have fins and enormous capacities for breath holding. Whales submerge up to two hours! The secret of breath holding is an amazing reduction of oxygen consumption. The diving mammals decrease it to as little as 15 percent the normal rate. They also decrease their heart rates, shunting most of the blood to the heart and brain.

Mammals adapt to the hottest and driest of climates as well as the coldest and wettest. Kangaroo rats and pocket mice, for instance, survive in the desert without drinking water, yet 65 percent of their body weight is water. Where does it originate? They get some from the "dry" plants they eat, and more as by-products of metabolism. (All animals produce metabolic water.) They conserve these small amounts of water by having no sweat glands, except on the feet, and by excreting little water in their urine and feces. Also, they live underground during the heat of the day, becoming active only at night.

Materials

- Snow, sand, or mud (outdoors)
- Guidebook to animal tracks

TRACKING MAMMALS

Lord Baden-Powell, the founder of the Boy Scouts, was a British general and renowned spy, unsurpassed at following animal or human trails and making appropriate deductions. Indeed, this was how he met his wife. Baden-Powell had learned to recognize character by a person's manner of walking. While in London one day he noticed the prints of a woman who "trod in a way that showed . . . honesty of purpose and common sense as well as the spirit of adventure."[1] Two years later he noticed the same gait in a fellow passenger sailing for the West Indies. Upon introducing himself, he established that this was the correct woman and eventually married her.

Experienced naturalists recognize in tracks many clues to animal activity—how they hunt or are hunted, where they live, and so forth. You can imitate these naturalists by repeatedly examining tracks in snow or in sand and mud along ponds and streams. Look for tracks near woods and in abandoned fields. Follow them. Identify them in a field guide, such as Murie's *A Field Guide to Animal Tracks*, and make life-size sketches. Was the animal moving slowly or swiftly? Fast runners leave prints far apart.

Two of the most frequently seen and easily recognized trails are those of cottontail rabbits and squirrels (FIGS. 13-1 and 13-2). Both animals leap. Their hind feet fall in front of their forefeet. The hind feet of squirrels point slightly outward. Toes and even toenails of the latter might be visible in fresh snow.

13-1   Rabbit trail

You can also learn about tracks by following domesticated animals. Take a pet for a walk through snow or mud. Observe that dogs and cats leave distinctly different trails when walking, trotting, and running. Notice that the nails of a cat, retracted while traveling, do not leave an imprint as do those of a dog.

Snow offers several clues about how recently an animal passed. Some melting occurs even on cold days. Fresh tracks are sharp and show clean, slick bottoms. Older tracks, especially those present for two or three days, become less distinct. During a continuous snowfall, tracks fill with snow.

Materials

- Tuna or chicken cans
- Petrolatum
- Plaster of Paris
- Water
- Small mixing bowl
- Knife

PREPARING CASTS OF TRACKS

To make a permanent collection of tracks, carry on hikes plaster of Paris and two or three tuna or chicken cans of different diameters. Before going, use a can opener to remove both ends of the cans, leaving circular metal bands. When you find a suitable footprint in mud or dirt, remove sticks and other debris that might have fallen into it. Smear a thin layer of petrolatum on the

Squirrel   Rabbit

Dog trot   Cat walk

13-2   Tracks of common mammals

inside of a metal band. Press the rim of the band into the soil around the track
(FIG. 13-3). Then mix some plaster of Paris with water to the consistency of
a thin pancake batter. As it just starts to thicken, pour it slowly into the
depressions of the track, continuing until the track and most of the circular
band are filled with plaster. Let the cast set and dry for 30 minutes or longer.
Then cut around and under the metal band with a knife to remove both it and
the cast from the soil. Rinse the cast with water and remove the metal band.
Label the cast with the date, location, and name of the animal that made the
track.

You can also make casts of footprints in snow. For this purpose, mix cold
water and snow with the plaster of Paris, getting the temperature of the mixture
very near that of the snow to prevent the footprints from melting.

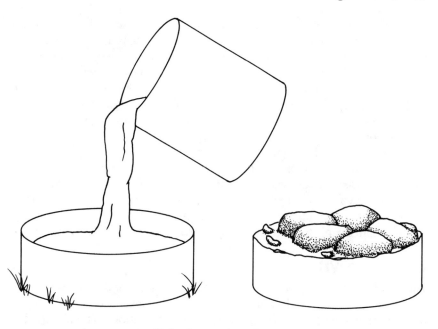

13-3   Preparation of a cast

Materials

- Binoculars
- Guidebook to mammals
- Camera

OBSERVING MAMMALS

Wherever you live, observe wildlife. Walk and look frequently, extending these walks, when possible, to parks or wilderness. Identify the mammals by using a guidebook, such as Burt and Grossenheider's *A Field Guide to Mammals*.

My family now lives on the Arizona desert. We often see pack rats, rabbits, coyotes, javelina, and other animals. The *pack rats* surround their burrows with segments of cactus, sticks, dried feces, and other debris, making it difficult for other animals to reach them. The *cottontails* and *jackrabbits* reproduce in the spring, as evidenced by many young. On hot days, they seek shade, and if the surrounding trees and rocks are cooler than their bodies, radiate heat to them from their large ears. The *coyotes* hunt and yelp mainly at night, giving us a warm feeling of oneness with nature. Sometimes they answer when we call. The piglike *javelina* travel in a wash (streambed) nearby. Our dog once chased about ten of them from our yard, but one circled around and began chasing the dog.

Many mammals are nocturnal, that is, active at night. To observe them closely, select one or more sites from which to watch by moonlight. Choose places where there are animal trails, for example, near water holes. Look for droppings, bits of fur, entrances to burrows, or other signs that the animals are nearby. If you wish, place food, such as salty peanuts, grain, or mealworms, on the trail or beside the burrows as bait. Go at dusk to a site downwind and off the trail. Camouflage yourself by wearing green or brown clothing, hiding behind bushes, and keeping quiet. Bring binoculars, preferably heavyweight binoculars that have large objective lenses; such lenses admit more light. Wear ample clothing for warmth and to keep off mosquitoes. Avoid mosquito repellents, the odors of which alert animals to your presence.

If you have a camera, take flash photographs. For animal photography, 35-millimeter, single-lens-reflex cameras are best. If you are far from your subject and if you have telephoto lenses, choose one having a focal length of perhaps 200 or 400 millimeters. Place the camera on a sturdy tripod, and use a remote control cable to snap the photograph.

OTHER ACTIVITIES

You can also study the behavior of domestic animals and animals at zoos. Among dogs, domestic or wild, watch for dominance and submission by observing the differences in facial expressions and positions of their tails. Also, watch for territorial marking and for the behavior of both males and females when the females are in heat. In zoos, watch monkeys grooming, apes using tools or toys, otters playing, seals swimming, and so on.

To study the range through which mice, voles, and other small mammals seek food, set trays of food near their burrows. Place 16 or more of these trays 5 to 10 meters (16 to 32 feet) apart along a checkerboard grid. Add bran, oats, or other grain to the trays daily. After a few days, color the grain in some of the trays, using different food colors for different trays. Mice and voles often defecate on the trays where they feed. By noting the color of the feces, you will see how far the animals traveled to feed.

Endnotes

1. Baden-Powell, R. S. S. *Lessons of a Lifetime*, p. 253. New York: Henry Holt and Company, 1933.

Part 4

Fearfully and wonderfully made

Chapter **14**

Anatomy of a mammal

"*I* am fearfully and wonderfully made," said the psalmist. "Marvelous are thy works; and that my soul knoweth right well."[1]

The body is made of *cells* that form four kinds of *tissues*: nervous, muscular, epithelial, and connective. Two or more tissues form *organs*, such as the brain and stomach. And a group of organs form *systems*, such as the nervous system and digestive system. Putting all systems together, we get the fearfully and wonderfully made body you are about to dissect.

Materials

- Preserved fetal pig or other mammal
- Dissection tray
- Surgical gloves
- Laboratory coat or old shirt
- Dissecting instruments, including forceps, sharp-pointed scissors, and scalpel

THE LANGUAGE OF ANATOMY

To be precise, we shall use anatomical terms to orient the dissection that follows.

Anterior refers to the front of a mammal, the part that moves forward. In a pig, for example, the anterior end is the head end. In a human, it is the side toward which the face and knees point.

Posterior refers oppositely to the tail end of most mammals and the back side of a human. In a pig, the posterior end is the tail end.

- *Dorsal* refers to the back side and *ventral* to the belly side of any vertebrate (animal with a backbone).
- *Midline* refers to the middle line running from the head to the tail. For example, an incision from the chin to the crotch runs along the midline.
- *Medial* means toward the midline. For example, one moves medially in going from the lungs to the heart.
- *Lateral* means toward the right or left sides. For example, one moves laterally in going from the heart to the lungs.
- *Right* and *left* refer to the animal's right and left sides, not yours.

DISSECTING THE HEAD, NECK, AND CHEST

These directions are for a fetal pig, but they also apply, with a few exceptions, to the fetuses and adults of other mammals. Before you dissect, thoroughly rinse the pig to remove excess preservative. As you dissect, wear surgical gloves and a laboratory coat or old shirt to keep the preservative off your hands and clothes. You can obtain intact animals and gloves from biological supply companies (Appendix B) or perhaps dissected animals from a college or high school laboratory. Drugstores often have surgical gloves.

CAUTION: Work outdoors or in a well-ventilated room. If you get preservative on your skin, wash it off several times with soap and running water. Then keep your hands away from your eyes.

When ready, place the pig with its ventral side up. Make a longitudinal, midline incision through the skin from the chin to the chest (FIG. 14-1). Extend the cut to the right and left at the upper and lower ends of the head and neck, and peel back the skin. It is bound loosely to the muscles by strands of connective tissue.

Look for the lumpy *salivary glands* at the base of the ears and near the jaw (FIG. 14-2). These glands secrete saliva into the mouth, lubricating the food and starting the digestion of carbohydrates.

Look also for a thin, longitudinal band of muscle (the sternohyoid) in the midline of the neck. Slit and remove the muscle to expose the lobular *thymus gland* beneath. The thymus extends from the neck into the chest, where it covers the upper end of the heart. The gland produces a kind of lymphocytes, called T cells. These cells provide immunity to viruses and fungi. In older animals, the thymus degenerates, but lymph nodes and other lymphoid organs continue to provide immunity.

Pull the lobes of the thymus aside to expose the dark, oval or triangular *thyroid gland*. The thyroid secretes thyroid hormones that stimulate heat production in mammals. When these hormones are deficient, animals become cold and sluggish.

Air comes through the mouth and nose to a muscular funnel, the *pharynx*, that leads into the *larynx* (voice box). The larynx is the sturdy, white, cartilaginous container of vocal cords, at the anterior end of the neck. The vocal cords vibrate to produce sound.

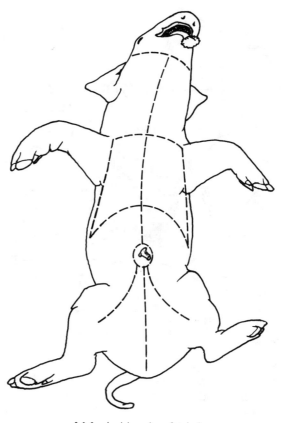

14-1   Incisions in a fetal pig

Air goes from the larynx to the trachea and its branches to the lungs. Expose the tubular *trachea* (windpipe). Tracheal rings of cartilage keep the passage open.

Pull the trachea slightly to one side to expose the *esophagus* beneath. The esophagus is the narrow, muscular tube that carries food from the pharynx to the stomach. Both air and food pass through the pharynx, allowing food occasionally to enter the airways or air to enter the stomach.

DISSECTING THE CHEST

Sever the skin along the midline of the chest and abdomen, extending the incision to the right and left at the junction of the chest and abdomen. Peel back the skin to expose the underlying muscles and *ribs*. Feel the ribs. Cut through the muscles and ribs along the two sides of the chest to fully expose the heart and lungs (FIGS. 14-1 and 14-2).

Find the *heart* near the midline of the chest. A thin-walled, translucent sack, called the *pericardium*, envelopes the heart. For a better view, cut away

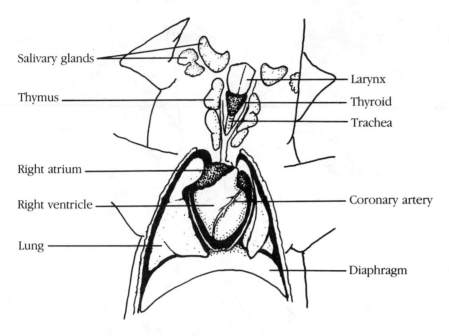

Salivary glands
Thymus
Right atrium
Right ventricle
Lung

Larynx
Thyroid
Trachea
Coronary artery
Diaphragm

14-2   Head, neck, and chest

most of the pericardium. The heart of mammals has four chambers: *right and left atria* at the anterior end and *right and left ventricles* at the posterior end. From your viewpoint, the atria are two dark flaps that extend slightly over the head end of the ventricles, and the ventricles appear as a single, blunt-ended chamber crossed by a *coronary artery*.

Trace the path of blood through the heart and lungs. It passes through anterior and posterior veins, called the *venae cavae*, into the *right atrium*. Then it goes to the *right ventricle, pulmonary artery*, and *lungs*. In fetal mammals, most of the blood in the pulmonary artery passes through a *ductus arteriosus* to the nearby aorta. At birth, the ductus arteriosus degenerates, causing blood then to pass only to the lungs.

Examine the lungs. They are soft and wet, like sponges. They are spongy because they contain many microscopic air sacs, called *alveoli*, and wet because blood is in capillaries surrounding the alveoli. After birth the pulmonary blood releases carbon dioxide into these alveoli and, in exchange, picks up oxygen. Because your pig is a fetus, its lungs have not yet filled with air.

Oxygenated blood travels through *pulmonary veins* to the *left atrium*. Then it goes to the *left ventricle, aorta*, and other *arteries*.

Cut out the heart and its large attached vessels (FIG. 14-3). Then cut into the cavities of the atria and ventricles. Compare the thickness of the muscle in the different chambers. That of the atria is thin because the atria pump blood weakly into the ventricles. Actually, most of the blood is drawn into the ventricles by suction as the ventricles relax between beats. Compare also the

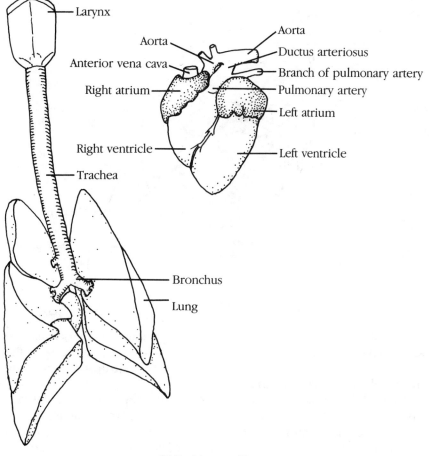

14-3   Heart and lungs

muscle of the right and left ventricles. After birth, that of the left ventricle is thicker because it pumps blood to all parts of the body except the lungs. That of the right ventricle is thinner because it pumps blood to the nearby lungs.

Return to the cavity of the chest. Notice how the trachea branches into *bronchi* (singular *bronchus*). The bronchi then branch repeatedly to carry air into the alveoli of the lungs.

At the posterior of the chest, between it and the abdomen, is a thin, domelike sheet of muscle called the *diaphragm*. When the diaphragm contracts after birth, the dome comes downward, drawing air through the trachea into the bronchi and lungs. The diaphragm causes inhalation.

DISSECTING THE ABDOMEN AND PELVIS

Cut along the midline of the abdomen through the skin and the thin layer of abdominal muscles (FIG. 14-1). Circle the *umbilical cord*, the tube through which the fetus gets oxygen and nutrients from its mother. Make incisions from

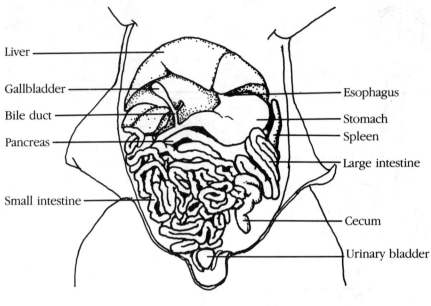

Liver

Gallbladder

Bile duct

Pancreas

Small intestine

Esophagus

Stomach

Spleen

Large intestine

Cecum

Urinary bladder

14-4   Abdomen and pelvis

the umbilical circle to the hind limbs. Turn back the skin and muscle flaps to see the abdominal and pelvic organs inside (FIG. 14-4). To see more, sever the umbilical vein, a vessel passing from the umbilical cord to the liver. Then pull the umbilical cord toward the tail.

Find the liver and gallbladder. The *liver* is the large, brown, multilobed organ curving under the diaphragm. It secretes bile, a green fluid that enters the intestine to break apart globules of fat. Between meals, the bile is stored in the *gallbladder*, tucked under the right side of the liver. The gallbladder has little color in fetal pigs.

Next find the esophagus and stomach. The *esophagus* is the muscular tube that passes through the neck, chest, and diaphragm, then enters the stomach. The *stomach* is the curved, muscular pouch located posterior to the diaphragm on the left side of the abdomen. After birth, contractions of the esophagus squeeze food into the stomach, where it is stored and mixed with gastric juice. Then contractions of the stomach squeeze it little by little into the small intestine. The stomach secretes pepsin, an enzyme that starts the digestion of protein.

The *small intestine* is the muscular tube that twists back and forth several times within the abdomen. Shortly below the stomach, it is joined by tiny ducts from the liver and pancreas.

The *pancreas* is the lumpy organ that curves along the rim of the stomach and first part of the small intestine. It secretes enzymes that enter the intestine to digest food, as well as hormones, such as insulin, that regulate the blood glucose.

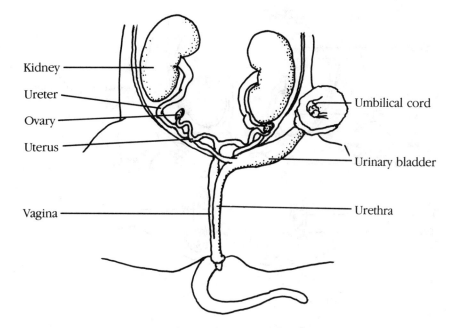

Kidney

Ureter

Ovary

Uterus

Vagina

Umbilical cord

Urinary bladder

Urethra

14-5 Urogenital tract of a female

Look for the *spleen*. It is a long, brown, slightly curved organ found next to the body wall along the left margin of the stomach. The spleen is a storage site for red blood cells, the carriers of oxygen. When the body needs more oxygen, the spleen contracts to expel its cells into the bloodstream.

Cut out a 2-centimeter segment of the small intestine, and slit it longitudinally. Feel the velvet-smooth interior. Microscopic, fingerlike villi provide the velvet touch. Nutrients are absorbed through the vast surface of the many villi.

After birth, contractions of the small intestine move food slowly downward, allowing time for digestion and absorption. The residue, mostly fiber, enters the *large intestine*, so named because its diameter is wider than that of the small intestine. In fetal pigs, the first part of the large intestine forms a compact, coiled mass. Then it passes posteriorly from the coil to discharge feces through the *anus* to the outside.

A fingerlike *cecum* projects posteriorly where the small intestine enters the large intestine. The cecum is the site in which bacteria digest the cellulose of plants.

Pull the intestines and cecum aside to locate the two, tan-colored *kidneys* on the dorsal wall of the abdominal cavity (FIG. 14-5). As blood enters the kidneys, some of its water, salts, and wastes become urine. The urine drains away from each kidney through a narrow tube, the *ureter*. Find the two ureters. They carry urine into a thin-walled *urinary bladder*, a pouch extending

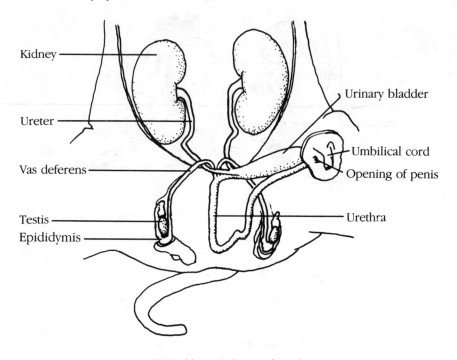

Kidney

Urinary bladder

Ureter

Umbilical cord

Vas deferens

Opening of penis

Testis

Urethra

Epididymis

14-6 Urogenital tract of a male

posteriorly from the umbilical cord to the pelvic cavity. This cavity, continuous with the abdominal cavity, is the space surrounded by the pelvic or hip bones.

The wall of the urinary bladder is muscular. When it contracts, it empties urine through a narrow, tubular passage called the *urethra*. Find the beginning of the urethra, the place where it connects to the bladder.

If you have a female pig, locate the tiny *ovaries* on the dorsal wall of the abdomen, posterior to the kidneys (FIG. 14-5). The ovaries are oblong structures that produce ova (eggs) and female hormones called estrogens. The ova later pass directly into tiny, coiled *uterine tubes* and from these into the *uterus*. Identify these structures. The uterine tubes are hidden on the dorsal sides of the ovaries.

In pigs, but not humans, the uterus branches, forming a tubular Y-shaped passage. In the anterior horns of the Y, fertilized eggs divide and develop into embryos. At birth, the resulting babies pass through the posterior tube of the uterus to the *vagina* and the world. To see the vagina and the central body of the uterus, cut through the bone at the midline of the pelvis. Then spread the legs of the pig, and look into the cleft.

If you have a male pig, find the opening of the penis immediately posterior to the umbilical cord (FIG. 14-6). Feel the cord-like penis under the skin. Then make a shallow, posterior cut throught the skin overlying the penis (FIG. 14-1). Extend the cut nearly to the tail. Find the two cream-colored, ovoid *testes* by peeling the skin laterally on both sides (FIG. 14-6). If the fetal pig is young, its testes might have descended only partway into the scrotum, the pouch that holds them.

The testes are the source of sperm and the male hormone, testosterone. The sperm are stored in an *epididymis* that curves longitudinally along the side of each testis. Eventually the sperm are ejaculated through a narrow, white tube, the *vas deferens*. The vas deferens connects to the urethra. Sperm or urine pass at different times through the urethra and out the *penis* that surrounds it.

DISSECTING THE BRAIN

Turn over the pig, putting its dorsal side up. Cut through the skin along the midline of its head and neck to expose the skull beneath. Then make lateral incisions through the skin, and peel it back from the skull. Use pointed scissors to puncture the thin bones, avoiding the brain inside. Then use forceps and scissors to break and pick away bits of bone, exposing the entire brain. (See chapter 16 for diagrams of the sheep brain, which is similar.)

The brain is covered by a thin, translucent membrane called the *dura mater*. You might have already cut the dura while entering the skull. If not, do it now to expose the cerebrum beneath. The *cerebrum* has right and left sections called *cerebral hemispheres*. These hemispheres detect and analyze sensations, such as touch, hearing, and vision, and direct bodily movements. The right hemisphere controls mainly the left side of the body and the left hemisphere controls mainly the right side.

Look at the posterior of the brain, just past the ends of the cerebral hemispheres. There you will see the *cerebellum*. If necessary, pick away more bone to expose it. The cerebellum works with the cerebrum to start and modify movements.

The brain of the fetus is soft and small, making it difficult to dissect. If you want to go further, see FIGS. 16-2, 16-3, and 16-4 of the sheep brain. Alternatively, dissect a sheep brain.

When you finish, remove your gloves. Wash the dissection tray, dissecting instruments, and your hands thoroughly with soap and water. If you want to keep the dissected pig, place it in a container of isopropyl alcohol.

OTHER ACTIVITIES

If you want to identify additional structures in your pig, such as the muscles and nerves, borrow a dissection manual from the library of a college or university. The library will likely have other manuals describing comparative anatomy. Using these, you can make comparative studies of a single organ, system, or all systems of invertebrates, fishes, amphibians, reptiles, birds, and mammals.

For further anatomical studies of the eye, muscles, and heart, see chapters 20, 25, and 30.

Endnotes

1. Psalms 139:14

Chapter 15

Constancy in our bodies

Our meals are widely spaced. When we eat, we digest and absorb food for hours, emptying the digestive tract; then no nutrients pass to the blood. Yet our brains keep thinking and our muscles keep contracting, so we must have a source of nutrients.

Delivery continues because nutrients—mainly *glucose*—are stored for release between meals. As we digest carbohydrates into glucose, the glucose enters the blood for transport to the liver. There the molecules of glucose combine, forming larger molecules of *glycogen* for storage. Hours later, during the fast between meals, this glycogen breaks down, releasing glucose into the blood for consumption by the cells. Whether eating or fasting, therefore, our glucose levels vary little—from about 120 milligrams per 100 milliliters of blood, to 80.

Our body temperatures also stay constant, usually between 36.5° and 37.5° C (97.7° and 99.5° F). We stay constantly warm because a thermostat in the brain—specifically in the *hypothalamus*—regulates our temperature as follows (FIG. 15-1):

1. When the temperature of the hypothalamus drops below its setting, the hypothalamus directs blood vessels in the skin to *constrict* (narrow in diameter). The constriction shifts warm blood to the interior of the body, reducing heat loss.

2. When instead the temperature of the hypothalamus goes above its setting, the hypothalamus directs blood vessels in the skin to *dilate* (expand in diameter). The dilation shifts warm blood to the skin, increasing heat loss.

Sometimes we expose ourselves to such extremes of temperature that

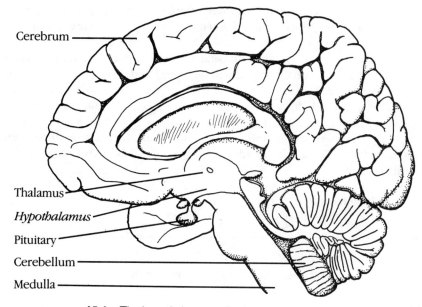

Cerebrum

Thalamus
Hypothalamus
Pituitary
Cerebellum
Medulla

15-1   The hypothalamus and other parts of the brain

changes in blood flow cannot fully compensate. In this case, the hypothalamus directs us to *sweat* or *shiver* or *move* to more comfortable surroundings.

Biologists use the word homeostasis to describe such stability. *Homeostasis* is the tendency of conditions in the body to stay constant. There are changes, to be sure, but they stay within narrow limits. If the usual limits of blood glucose, body temperature, and other functions are exceeded, discomfort or death can follow.

Materials

- White and colored chalk
- Chalkboard
- An assistant

HOW FEEDBACK WORKS

In the body, *feedback* helps us maintain homeostasis. For example, when a person sweats in response to heat, the evaporation of sweat cools the blood. The cooler blood feeds back to the hypothalamus, causing it to adjust the secretion of sweat. Thereby the temperature of the blood is kept constant.

In this experiment, you will use feedback from another person to help you draw a constantly straight line. Begin by looking as you draw a first straight line across a long chalkboard. Then place chalk of another color at the beginning of the line, and turn so that you cannot see it. Ask your helper to start clapping his or her hands at a steady rate of about twice per second. Say that you will attempt to draw a second line on top of the first. Ask the helper to clap faster if you move above the original line and slower if you move below it. Thereby you will have feedback to guide you.

As your helper claps, start drawing the line steadily across the board without looking. As you draw, you will hear the handclaps getting faster and slower to guide you.

When you finish, look at the board. Your second line alternately moves above and below the original line (FIG. 15-2). This is how homeostasis works. Your body temperature, blood glucose, and other functions oscillate from normal settings. Feedback provides the signals that help you to correct the oscillations.

Using feedback

15-2   Homeostasis shown by drawing chalk lines

Materials

- Oral or rectal thermometer and temperature band
- Hot lemonade, hot cider, or hot water
- Sherbet or crushed ice

STABILIZING THE BODY TEMPERATURE

The temperature of the body varies with the site at which we take it. In humans, the temperatures of the hypothalamus and mouth are about 37° C (98.6° F) and of the armpit about 36.5° C (97.7° F). Those of the forehead, hand, and fingers, respectively, are about 34.5° C (94.1° F), 33° C (91.4° F), and 32° C (89.6° F).

The temperature of the environment greatly affects these averages, especially in the skin of the extremities. Constriction and dilation of blood vessels in the fingers and toes change the volume of blood flowing through them up to 200-fold. In addition, the fingers and toes are too small to retain heat.

You can cause changes in blood flow and skin temperature by drinking large volumes of hot or cold fluids, beginning with hot. This will allow you also to feel sweating. Start this experiment when your body feels comfortably neutral in temperature.

Begin by measuring and recording the temperatures of your armpit and index finger, and if you wish, other parts of the body. For the armpit, press the tip of an oral or rectal thermometer deep into it, then hold your arm against your chest for three minutes. For the finger, hold a temperature band against it for the time recommended by the manufacturer (FIG. 15-3).[1]

Now drink 1 liter (1 quart) of hot lemonade, hot cider, or hot water quickly. Its temperature should be slightly cooler than that of hot coffee or tea, allowing you to drink it faster. The hot liquid enters your stomach within

15-3　Temperature band showing a skin temperature of 32° C

seconds, where it heats the blood of surrounding vessels. The heated blood, diluted by other blood, goes throughout the body, raising the hypothalamic temperature about 0.5° C (0.9° F).

After you drink about half the hot liquid, measure and record your finger temperature as you continue drinking. When finished drinking, immediately measure and record the temperatures of your armpit and finger. Also observe the amount of sweat. Measure and record the temperatures and observe the sweat again at 10 and 20 minutes after drinking.

How much greater is the increase in temperature of the finger than of the armpit? How is vessel dilation linked to the temperature increase in the skin of the finger? How is the hypothalamus related to the dilation of vessels and the increased rate of sweating? And how do these effects help stabilize the body temperature?

Although the temperature of the skin rises as you drink the hot liquid, the rise might stop or the skin might cool when the sweat glands secrete sweat. Evaporation of sweat cools the skin and the blood flowing through it. This is why the body needs a deeply placed thermostat—that of the hypothalamus—to regulate its temperature. The sensors of the skin sometimes give the wrong information about heat.

When the body again feels comfortably neutral or just slightly warm, remeasure and record the temperature of the armpit and finger. Then quickly eat 1 liter (1 quart) of sherbet or crushed ice, or as much as you feel comfortable in eating. As the ice melts in the mouth and stomach, it chills the blood of surrounding vessels. The cool blood then circulates throughout the body, causing reductions in body temperature.

After you eat about half the ice, measure and record your finger temperature as you continue eating. When finished with the ice, immediately measure and record the temperatures of both your armpit and finger. Also observe any increase in muscle tone—that is, a tightening or stiffness of muscles—the forerunner of shivering. Measure and record the temperatures and observe the tone again after 10 and 20 minutes.

Does the temperature of the finger decrease more than that of the armpit? How is vessel constriction linked to the temperature decrease of the finger? How is the hypothalamus related to the constriction of vessels and increase in muscle tone? How do these effects help stabilize the body temperature?

Weak contractions of muscles increase their tone, and strong contractions cause shivering. Both mechanisms increase heat production. Vigorous shivering generates as much heat as a brisk walk.

Materials

- White stationery
- Betadine (iodine) swabs or cotton-tipped swabs soaked in Lugol's (iodine) solution
- Transparent or adhesive tape
- Hot lemonade, hot cider, or hot water

LOCATING THE ACTIVE SWEAT GLANDS

You can study the sweat response more fully by locating the active sweat glands. To do this, paint two or more 1.5 centimeter square areas of your skin with a Betadine (povidone-iodine) swab or a cotton-tipped swab soaked in Lugol's solution. I suggest painting your wrist and forehead.

Now cut several 1-centimeter squares of white stationery. When the painted areas are thoroughly dry, use transparent or adhesive tape to bind the squares of white paper to the painted skin. After 30 minutes, remove the papers to see the blue- black dots that represent the positions of active sweat glands. The iodine in the pores dissolves in sweat and reacts with starch in the paper, producing the dark spots (FIG. 15-4).

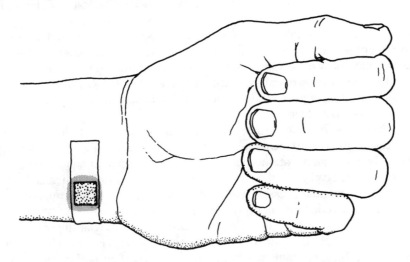

15-4 Location of sweat glands

Repaint the skin at adjacent sites, tape on new squares of white stationery, and drink 1 liter (or 1 quart) of hot lemonade, hot cider, or hot water. Wait again for 30 minutes, and remove the papers. Did the hypothalamus activate more sweat glands to cool the body? Was more sweat released by each gland?

STABILIZING THE BODY'S WATER CONTENT

Did you notice that when you drank 1 liter of water, you excreted a large volume of urine? When you drink little water, you excrete little urine and become thirsty. These are the body's means for stabilizing its water content and volume of blood. The center of control is in the hypothalamus of the brain.

The hypothalamus detects and acts upon thirst, deciding whether you drink. When you have little water in your blood, you feel thirsty and drink to replace the missing water. When you have excessive water in your blood, you feel hydrated and do not drink.

The hypothalamus also regulates urination, deciding the volume of urine you excrete. When you have little water in your blood, the hypothalamus makes an *antidiuretic hormone*. This hormone acts on your kidneys, causing them to retain water rather than excrete it. In contrast, when you have excessive water in your blood, the hypothalamus stops making the antidiuretic hormone. The lack of this hormone causes the kidneys to excrete more water in urine.

OTHER ACTIVITIES

The body temperature remains almost but not entirely constant. People have daily rhythms in which their temperatures rise to a peak, usually in the afternoon, then subside. In women, the temperature also varies with the menstrual cycle. It increases about 0.3° C (0.5° F) during the two weeks between ovulation (egg release) and the menses (bleeding). You can find the shifts in your own temperature by taking oral temperatures every hour for one day and, if you are a woman, every day for one menstrual cycle of approximately one month. If you take daily temperatures, measure them when you first awake each morning.

When you next develop a fever, notice how it progresses. Bacterial and viral toxins reset your hypothalamic thermostat. In response, your blood vessels constrict in the skin, your muscle tone increases or you shiver, and you put on more clothing or seek warmer surroundings. By these means you retain and generate more heat, causing a fever that allows the body to reach the new setting. The temperature stabilizes at its higher level—a temperature at which you are comfortable—until the toxins decrease. Then the thermostat is reset at a lower temperature. In response, your vessels dilate, you sweat, and you remove clothing or seek cooler surroundings. By these means you dispose of heat and reach the lower setting.

Endnotes

1. You can buy temperature bands (flexible strips) from medical and biological supply companies and some drug stores. Manufacturers sometimes calibrate the bands to show temperatures higher than the real temperatures, making the forehead and oral temperatures appear equal. To find the calibration, read the manufacturer's literature.

Part 5

Nerves and Brain

Chapter **16**

Anatomy of
a brain

*T*o learn how the brain works, biologists sometimes study the effects of strokes. Strokes occur when clots or bleeding stop blood flow to parts of the brain, causing death of its cells, interfering with activities controlled by these cells. When strokes destroy cells in the *frontal lobes* (front lobes) of the cerebrum, for example, the victims can no longer move certain parts of their bodies. Their arms, legs, or speech apparatus are paralyzed. From this fact we learn that the frontal lobes control movement.

Other strokes, wounds, or tumors destroy cells in the *parietal lobes* of the cerebrum, the lobes directly behind the frontal lobes. The victims might then forget parts of their bodies because they cannot feel them. The parietal lobes normally analyze touch and position. Dr. J. M. Nielsen describes one of his patients with parietal damage:

> He bathed the right side of the body and dressed the right side only. When his wife called attention to the discrepancy he recalled that he had left limbs and took care of them. During this process of bathing and dressing he used the left limbs correctly. . .but. . .was not conscious of them; he gave them no attention.[1]

Nielsen describes other patients who forgot their limbs entirely or disowned them. He asked one patient to follow her limbs visually from the trunk outward. "But my eyes and feelings don't agree," she said, "and I *must* believe my feelings. I know they look like mine, but I can feel they are not, and I can't believe my eyes."[2]

Stranger yet is the result of damage to the *occipital lobes*, the lobes at the hind end of the cerebrum. The occipital lobes normally recognize what the eyes see, but if damaged, they do not. Dr. Oliver Sacks describes a patient so defective that he could not tell his foot from his shoe nor the heads of children from those of fire hydrants. When Sacks finished an examination, the patient "started to look round for his hat. He reached out his hand, and took hold of his wife's head, tried to lift it off, to put it on. He had apparently mistaken his wife for a hat!"[3]

You can read these and other case histories by getting the books listed in the endnotes. Meanwhile, dissect a sheep brain or other mammalian brain, as here described, to get better acquainted with its parts. The names and functions of these parts are mostly the same as in humans, though our cerebrums are larger and our thoughts more profound. As you dissect, imagine what damage to the brain would do to behavior.

Materials

- Sheep brain or other mammalian brain
- Surgical gloves
- Dissection tray
- Dissecting instruments (scissors, forceps, and knife or scalpel)

EXAMINING THE OUTSIDE OF THE BRAIN

You can use any mammalian brain, fresh or preserved, for this dissection, but the directions are written specifically for a sheep's brain. Buy a brain that has its pituitary attached (Appendix B). For fresh brains, contact a custom butcher or meat packer listed in the yellow pages of your phone book. If you dissect the brain of a mammal other than a sheep, its parts will be larger or smaller than here described, but all parts will be present.

During the dissection, wear surgical gloves such as those worn by dentists. The gloves keep the preservative or other matter off your hands.

CAUTION: If you get preservative on your skin, wash it off several times with soap and running water. Then keep your hands away from your eyes.

Begin by removing the brain from the skull or its package. If it is preserved, rinse it thoroughly with water, flushing most of the irritants that might get on your skin or eyes. Then put the brain in a dissection pan.

Compare the brain with the drawing of a sheep brain (FIG. 16-1). Get oriented. Your sheep brain, unlike the one in the drawing, might still have its *meninges* intact. The meninges are thin, translucent, sheet-like coverings of the brain and spinal cord. Sometimes they get inflamed, a condition called meningitis. The inflammation results usually from a transfer of bacteria or viruses through the blood or an open wound. The bacteria or viruses cause fevers and headaches.

If the brain has meninges, look at them and remove them. Thereby you will see the underlying parts of the brain. Also remove excess fat, bone, and connective tissue under the frontal lobes at the anterior (front) end of the brain. This is the place where the eyes connect through *optic nerves* to the brain. The eyes were probably cut out before you purchased the brain, but the optic nerves will still be present. Save the pituitary for later dissection. It lies under the brain and just behind the optic nerves.

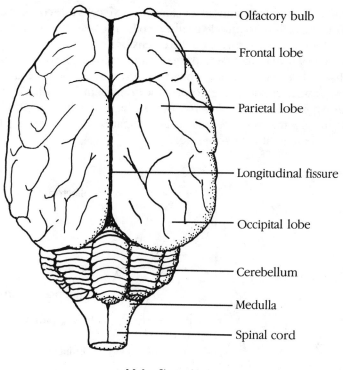

Olfactory bulb

Frontal lobe

Parietal lobe

Longitudinal fissure

Occipital lobe

Cerebellum

Medulla

Spinal cord

16-1 Sheep brain

Place the brain top side up in the dissection pan. Notice that there are two *cerebral hemispheres* separated by a deep groove called the *longitudinal fissure*. At the anterior or front end of the hemispheres are the *frontal lobes*. In sheep, humans, and other mammals, the frontal lobes help regulate bodily movement. When the movement area of the frontal lobes is damaged—as by a stroke—the victims find themselves partially paralyzed.

Going posteriorly (backward) from the frontal lobes of the cerebrum, we come first to the *parietal lobes*, which analyze touch and position, and then to the *occipital lobes*, which analyze sight. These lobes of the brain and the temporal lobes are harder to distinguish in sheep than in humans. The *temporal lobes* lie under the temple and ears, ready to analyze what we hear. Damage to the temporal lobes or the hearing apparatus of the ears, interferes with hearing.

The surface of the cerebrum is *convoluted*—that is, formed by hump-backed ridges. Make a V-shaped cut about 1 centimeter (½ inch) deep into a convolution. The first few millimeters consist of dirty-white *gray matter* and the rest of lighter-colored *white matter*.

The gray matter, also called the *cerebral cortex*, contains invisibly small *nerve cell bodies*. These are the parts of nerve cells that direct movements, detect and analyze sensations, and think. The convolutions are more numerous and deep in humans than in sheep, giving us more cell bodies with which to think.

The white matter contains extensions of nerve cells that connect each cell to other nerve cells or to muscle or gland cells. The matter is white because it contains lipids (fats) that surround the extensions.

Turn now to the underside of the brain (FIG. 16-2). Notice the *olfactory bulbs* that connect through minute olfactory nerves to the nasal passages. These nerves were torn free during removal of the brain from the skull. The olfactory bulbs lead posteriorly to the *rhinencephalon*, the part of the brain that analyzes odor. In humans the bulbs are quite small compared to those of sheep because we depend little on the sense of smell.

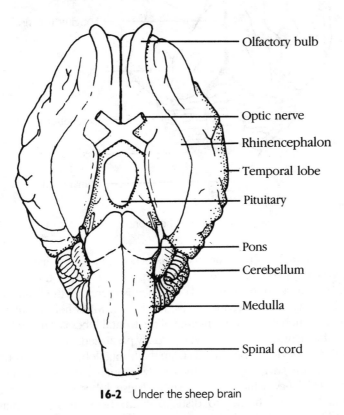

Olfactory bulb

Optic nerve

Rhinencephalon

Temporal lobe

Pituitary

Pons

Cerebellum

Medulla

Spinal cord

16-2 Under the sheep brain

Notice again the *optic nerves* leading from the eyeballs into the brain. The optical connections in the brain terminate in the occipital lobes, the analyzers of sight.

Other nerves lead also to and from the brain. You might notice a few of these entering the underside of the brain, but most were torn loose in removing the brain from the skull. We shall not identify those remaining.

Turn again to the occipital lobe. Look behind it. Here you will find a large, tightly convoluted structure called the *cerebellum*. It works with the frontal lobes to coordinate movements. When the cerebellum is sufficiently damaged, the victim staggers like a drunk.

Move again to the underside of the brain. Note the pituitary, which we will consider later. Posterior to (behind) the pituitary, you will see a rounded eminence, the *pons*, then an elongated eminence, the *medulla*, and finally a portion of the *spinal cord*. The pons links the upper and lower parts of the brain and moderates breathing. The medulla is of greater concern because it directs breathing and helps direct blood circulation. If the medulla is damaged sufficiently, breathing stops. The spinal cord is also crucial because it joins most of the nerves from the brain to the body. When the spinal cord is severed, the victim becomes paralyzed and cannot feel sensations below the cut.

EXAMINING THE INSIDE OF THE BRAIN

Locate again the longitudinal fissure, the depression that separates the two hemispheres. Using a sharp knife or scalpel, cut downward through this fissure to produce two halves of the brain, each half being a duplicate of the other (FIG. 16-3). In cutting, your knife first penetrates the *corpus callosum*, a longitudinal band of nerve-cell branches that connect the two cerebral hemispheres. The corpus callosum allows the two sides of the brain and the body to operate as a single, coordinated unit.

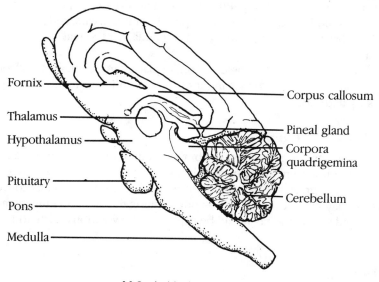

16-3   Inside the sheep brain

The *fornix*, another band of nerve-cell branches, lies under the corpus callosum at an acute angle to it. The fornix connects some of the deeper parts of the brain.

The *thalamus* lies below the corpus callosum and fornix. All sensory neurons, except those for smell, pass through the thalamus to the sensory parts of the cerebral cortex. Damage to the thalamus, therefore, can interfere with the reception of many sensations, including those for touch, hearing, and sight.

The *hypothalamus* lies below the thalamus, "hypo" meaning under. The hypothalamus helps regulate emotions, feeding, drinking, urination, body temperature, and hormone secretion. If an investigator, for example, continuously stimulates the drinking region of the hypothalamus with a gentle, imperceptibly small voltage, the animal will start drinking and continue drinking until it consumes half or more of its daily intake of water.

Nerve cells and blood vessels from the hypothalamus connect into the *pituitary gland* that lies underneath. The nerves and vessels transfer hypothalamic hormones to the pituitary—hormones that direct the release of pituitary hormones. The pituitary hormones, in turn, circulate throughout the body, where they direct growth, reproduction, and metabolism. If the hypothalamus or pituitary produces too little or too much of the hormones, the hormones change the operation of the body. For example, if the glands produce too much growth hormone, children grow seven, eight, or nine feet tall.

Look now for the *pineal gland*, shaped like a tiny pine cone, that projects above and behind the thalamus. The pineal cooperates with the testes and ovaries in regulating reproduction.

Directly posterior to the thalamus is the *midbrain* that includes four bumps called the *corpora quadrigemina*. The midbrain contains nerve tracts that connect parts of the brain, and the corpora quadrigemina perceive moving sights and sounds. They help us follow people, animals, cars, and so on.

This brings us back to the pons, medulla, and spinal cord, already identified.

If you want to keep the dissected brain, place it in a container of alcohol. Wash the dissection tray, dissecting instruments, and your hands thoroughly with soap and water.

OTHER ACTIVITIES

You can compare the anatomy of the brains of different animals. If you do, you will see the brain enlarging from fishes to amphibians, reptiles, birds, and mammals. The greatest enlargement occurs in the cerebrum and its convolutions. The latter are deepest and most numerous in the brains of humans, apes, whales, and porpoises—as would be expected. Convolutions relate to intelligence.

Endnotes

1. Nielsen, J. M. *Agnosia, Apraxia, Aphasia*, 2nd ed., p. 77. New York: Paul B. Hoeber, 1946.

2. Nielsen, p. 78.

3. Sacks, Oliver. *The Man Who Mistook His Wife for a Hat and Other Clinical Tales*, pp. 10 and 12-13. New York: Summit Books, 1985.

Chapter **17**

Reflexes
and reactions

*R*_{eflexes} are subconscious responses to stimuli, guided by the brain or spinal cord. Our eyes, for example, reflexly blink when an object flies toward them. The sight of the object activates movement-control areas of the brain, those that direct muscles to close the eyelids. Similarly, our stomachs reflexly secrete digestive juices as we chew food. The smell and taste of food activate nerve cells of the brain that cause the gastric glands to secrete. Also, our knees reflexly jerk when a physician taps the patellar ligament. The tap activates the ligament, spinal cord, and muscles that cause the jerk.

Though reflexes are simple events, they often follow one another in complex chains. A frog, for example, upon seeing a fly, turns its head, flips out and retracts its tongue, swallows the fly, and secretes digestive enzymes in an unvarying reflex chain. It seems to act thoughtfully, but the responses are subconscious and machinelike. They resemble the responses of a vending machine for iced drinks.

To illustrate another reflex in detail, let us assume that you are barefooted and step on a nail. In a fraction of a second—before you understand what happened—you withdraw your foot. Biologists call this a *withdrawal reflex*. Why do you withdraw your foot? You withdraw it because the stimulus, a nail, excites pain receptors. These particular receptors are bare nerve endings. When hit by a nail, they send impulses—low-voltage electrical signals— through *sensory nerve cells* to the spinal cord. In the cord, the impulses transfer to *connecting nerve cells* and from the connectors to *motor nerve cells*, the ones that direct movement. The motor cells lead to muscles of the leg which contract, pulling the foot away from the nail (FIG. 17-1).

Reflexes are subconscious responses. A brainless animal—one from which the brain is surgically removed—still has spinal reflexes, though it cannot feel them. Moreover, some reflexes are unfelt even with the brain intact.

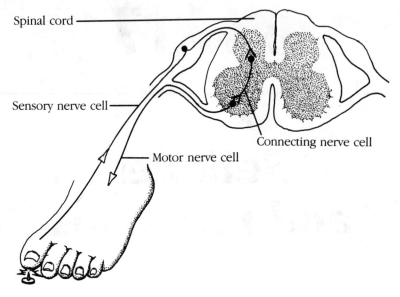

Spinal cord

Sensory nerve cell

Connecting nerve cell

Motor nerve cell

17-1   Reflex response to pain

How many of us, for example, feel our stomachs secrete digestive juices in response to food? But all of us know that a nail in the foot hurts and that more than our feet and legs respond. The message from the nail passes to the conscious brain as well as the subconscious spinal cord. The brain, including its speech-making parts, is aroused, as evidenced by the outcry. But the cry is unneeded for the subconscious withdrawal of the foot.

In contrast to reflexes, *reactions* are conscious responses to stimuli. A light turns green, you cross the street. A car comes near; you move to avoid it. In each case the mind recognizes a stimulus and consciously produces a response.

We shall study both subconscious reflexes and conscious reactions in the experiments that follow.

Materials

- Human subject
- Goggles or face-sized sheet of transparent plastic
- Cotton balls or paper wads

EYE-BLINK REFLEX

Ask a subject to put on goggles or to hold a sheet of transparent plastic in front of his or her eyes. Then toss cotton balls or paper wads toward the protected eyes. Does the subject subconsciously blink as the objects come near the eyes? Can the subject consciously resist blinking, that is, can the conscious brain suppress the subconscious brain?

Materials

- Mirror
- Dark room
- Source of light

PUPIL REFLEX

With a light turned on at night, look in a mirror to see the pupils of your eyes. If the room is bright, the pupils will appear small. Now turn off the light for one or two minutes; then switch it on as you look again at the mirror. Darkness causes the pupils to dilate (enlarge), and light causes them to constrict (narrow).

Dilation is a reflex that admits more light to the eyes. The radial muscles in the iris—the colored part of each eye—contract, pulling the rim of the iris outward. Conversely, constriction is a reflex that admits less light to the eyes. The circular muscles in the iris contract, drawing the rim of the iris inward.

Materials

- Pet dog

SCRATCH REFLEX

Vigorously tickle the chest or abdomen of a friendly dog. The irritation of the tickle or of fleas on its skin reflexly activates nerve cells in its spinal cord. These cells direct kicking of its hind leg.

Materials

- Lemon juice

SALIVATION REFLEX

If you have a pet dog, you might have noticed it salivate or lick its lips when you get its food. Perhaps you have even noticed your own mouth watering as you smell and contemplate good food. Such salivation is a reflex, a subconscious response to smell and taste. The receptors for smell and taste act through sensory neurons, the brain, and motor neurons to activate the salivary glands.

Sour tastes seem especially effective. Place 2 or 3 drops of lemon juice on your tongue, close your mouth, and hold it closed for 1 or 2 minutes. Do you feel the saliva collecting?

Materials

- Human subject
- Cup of water
- A stethoscope and isopropyl (rubbing) alcohol (optional)

SPHINCTER REFLEX

A *sphincter* is a circular band of muscle that contracts or relaxes to close or open a circular opening. You have already seen the operation of one sphincter, that of the iris that regulates the closing and opening of the pupil. Now you will study another sphincter, the one at the lower end of the esophagus that regulates the passage of food and water into the stomach.

Wash the earplugs of a stethoscope with isopropyl (rubbing) alcohol, and insert them into your ears. Then place the disk of the stethoscope or, if you have no stethoscope, an ear against the abdominal wall of a subject, slightly to the left of the abdominal midline and at the base of the ribs. Have the subject swallow water. Listen! In 5 to 10 seconds, you will hear gurgling. The gurgling results from a reflexive relaxation of the sphincter. Water now enters the stomach.

Materials

- Human subject
- Chair

KNEE-JERK REFLEX

Using the side of the hand, strike just below the patella (kneecap) of another person sitting with crossed legs. The leg jerks. In this reflex, the *patellar ligament* over the kneecap is stretched. Receptors for stretch direct impulses through the spinal cord, causing the contraction of muscles in the thigh.

Physicians check the knee jerk and other reflexes to find the condition of the neuromuscular system. If there is no response or an exaggerated response, they suspect damage. By testing reflexes in different parts of the body, they find the location of the damage.

ANKLE-JERK REFLEX

Have your subject stand beside a chair, with one leg kneeling on the seat. Strike the tendon of Achilles that connects the calf muscles to the heel. (You can feel this tendon at the back of the ankle.) The foot jerks backward. Receptors in the tendon and muscles respond to the sudden stretch, causing the muscles to contract reflexively.

POSTURE REFLEX

Ask your subject to stand quietly as you feel his or her calf muscles. They are weakly contracted. Now ask the subject to lean forward as far as possible without falling. Notice that the muscles contract strongly to prevent the fall. The contraction results from a stretch reflex similar to that obtained with the ankle jerk, but more prolonged. The stretch receptors are located in the calf muscles and tendon of Achilles.

ABDOMINAL REFLEX

Rapidly draw your finger across the subject's abdomen just beneath the bottom ribs. Abdominal muscles contract, causing the abdomen to draw in.

Materials

- Human subject
- Ruler, yardstick, or meterstick

REACTION TIME TO A FALLING RULER

It is difficult in most home or school laboratories to measure reaction time in fractions of a second. It is easy, however, to measure the distance through which an object falls, and from this to calculate the time. To do this, hold a ruler by the 12-inch mark. Ask another person (the subject) to place his or her thumb and first finger over, but not touching, the 0-inch mark (FIG. 17-2). Keep the fingers about 1 inch from the ruler. Release the ruler, and have the subject grab it as soon as he or she sees it fall. Note the distance of the fall, and compare it with TABLE 17-1 or 17-2 to find the reaction time. Have the subject also test you.

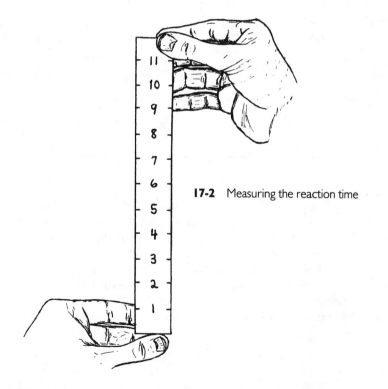

17-2 Measuring the reaction time

Usually the ruler falls 6 to 10 inches before it is caught, but if you divert the subject's attention, it will fall farther. Have the subject do mental arithmetic or

Table 17-1   Reaction Time Obtained from Distance of Fall in Inches

Inches	4	5	6	7	8	9	10	11
Seconds	0.14	0.16	0.18	0.19	0.20	0.22	0.23	0.24
Inches	12	13	14	15	16	17	18	19
Seconds	0.25	0.26	0.27	0.28	0.29	0.30	0.31	0.31
Inches	20	21	22	23	24	25	26	27
Seconds	0.32	0.33	0.34	0.35	0.35	0.36	0.37	0.38

For additional calculations use the formula:

$$\text{Time in seconds} = \sqrt{\frac{\text{fall in inches}}{192}}$$

Table 17-2   Reaction Time Obtained from Distance of Fall in Centimeters

Centimeters	10	12	14	16	18	20	22	24
Seconds	0.14	0.16	0.17	0.18	0.19	0.20	0.21	0.22
Centimeters	26	28	30	32	34	36	38	40
Seconds	0.23	0.24	0.25	0.26	0.26	0.27	0.28	0.29
Centimeters	42	44	46	48	50	52	54	56
Seconds	0.29	0.30	0.31	0.31	0.32	0.33	0.33	0.34

For additional calculations use the formula:

$$\text{Time in seconds} = \sqrt{\frac{\text{fall in centimeters}}{488}}$$

have someone run a hand unexpectedly through the subject's hair. During the diversion, drop the ruler or, better yet, a yardstick or meterstick, and determine whether the subject's reaction time is prolonged. If you try mental arithmetic, use the average of several trials.

Now ask the subject only to catch the ruler or other stick when you say a specific one-syllable word, such as "catch," as you drop it. Sometimes say catch and other times say different words, such as "fetch" or "itch." Measure the reaction time when the subject responds to the correct word but not the others. What happens and why?

Materials
- Human subject
- Dollar bill

REACTION TIME TO A FALLING DOLLAR

A variation of the previous experiment is to bet someone that he or she cannot catch a dollar bill if you drop it. Have the person place the thumb and first finger over, but not touching, the picture of George Washington. Drop the bill. You will not likely lose it.

REACTION TIME TO HAND SLAPPING

Hold out both of your hands, palms up, in front of your waist. Have the subject place both of his or her hands, palms down, on top of yours. Now try to slap the tops of the subject's hands before the subject can withdraw them. Does the subject react quickly enough to avoid the slap? Then reverse the procedure, having the subject attempt to slap your hands.

OTHER ACTIVITIES

To which would a person react faster: strong stimuli or weak stimuli? To find out, try the following. Paint one of two rulers a bright, fluorescent color and the second the color of the nearby walls that serve as a background. Measure the reaction times as a subject catches first one ruler, then the other. Alternatively, catch one ruler in a brightly lighted room and then in a dimly lighted room. Strong, well-defined stimuli excite more nerve cells more often, better alerting the brain. The alert brain responds more quickly.

Chapter **18**

Learning

*H*ave you noticed how excited dogs get at mealtimes? As their owners go for the food, the dogs jump about, their mouths water, and their tails wag. But when puppies first see their owners seek food, they make no response. Only when the puppies associate the movement with the food do they learn to anticipate. Experience conditions their behavior.

Ivan Pavlov, a Russian physiologist, was the first to study this *conditioned response* in detail. He experimented with dogs, determining how much saliva they released when he brought food nearby. Next he rang a bell, then fed them. After several trials, the animals answered the bell by salivating, even when he withheld the food.

Mazes provide another means to test learning. You have probably drawn your way through the alternative pathways of a printed maze. You started at one point and got to another, trying not to cross lines or enter blind alleys. Humans have little trouble with such problems, not because of superior intellect but because they use their eyes before they use their pencils. To make mazes more difficult for humans, experimenters sometimes build human-sized corridors.

Ants and worms become increasingly accurate as they grope along mazes, but mice and rats learn faster. Indeed, rats sometimes make fewer mistakes than humans! Ordinary mazes do not distinguish the intellect of animals more highly evolved than rodents. To show differences, special mazes are built which call for alternating right with left turns. Usually the animal makes two right turns during the first trial to get its reward. Next it makes two left turns for the same reward, then two right turns, and so on. Rats fail to solve this problem after a thousand attempts. Monkeys and human children do poorly, but human adults do well.

Another test of comprehension requires *delayed responses*. For example, food is placed under a triangular but not a circular piece of wood. Hungry, observant animals immediately remove the triangle and start eating. If taken from the room, however, rats forget where the food is. In contrast, monkeys remember for some time and humans almost indefinitely.

The true mark of intellect is reason or insight. A chimpanzee placed in a cage with two jointed sticks—neither of which is long enough to reach a banana outside—might poke around, then suddenly fit the sticks together to draw in its prize. It "sees" the solution. Occasionally monkeys and chimpanzees see solutions overlooked by their keepers. A psychologist once dangled a banana from the ceiling and put several boxes on the floor near his chimpanzee. The animal, disregarding the boxes, led the psychologist under the banana, then scrambled up the man's body to reach it.

Materials

- Dog
- Cat or other pet
- Meat or other food
- 2 meters of wire fence

HOW BRIGHT IS YOUR PET?

If you have a dog, cat, or other reasonably intelligent pet, try this simple test. Stretch a piece of wire fence, about 2 meters (6 feet) long, directly in front of your pet. Drop a favorite morsel of food behind the fence (FIG. 18-1). What is the reaction? A young puppy usually scratches and barks for some time before wandering to the other side. A chicken takes much longer. Older, more experienced dogs might go around at once, perhaps doubting your intelligence for dropping the food on the wrong side. If your pet does not immediately solve the problem, repeat the test two, three, or more times to find how quickly it learns.

Materials

- Mirror
- Watch with a second hand

HOW FAST DO YOU LEARN?

Pick a short paragraph in this book. Hold a mirror in front of the page and begin reading the mirror image aloud as quickly as possible (FIG. 18-2). Note the duration of reading from start to finish. Then reread the same paragraph. Are you faster? Repeat the reading several times for maximum improvement. Also read a different paragraph in the mirror. Does your previous training help?

In this test you became familiar with reversed print as well as the actual words of the paragraph. Reading aloud reinforced your learning. As impulses passed repeatedly through a circuit of nerve cells in your brain, the impulses temporarily stimulated the production of chemical transmitters, the transmit-

18-1   Will she get the hot dog?

ters that transfer impulses between nerve cells. Thus it became easier to arouse this circuit of thought.

Wait several days and again read the same passage. You will take less time than when you first read it but more time than when you last read it. As weeks and months pass, your recall decreases.

Repetition fixes knowledge in your mind by several means, including the growth of new connections between nerve cells. Branches of these cells grow outward like the branches of closely spaced trees, making contacts through which to pass memories. In experimental animals, such branches have been shown to increase the weight of the affected part of the brain. If the connections go unused, they and their memories wither.

Materials

- Human subject
- Pencil or pen
- Index cards or paper
- Watch with a second hand

LEARNING SENSE AND NONSENSE

We learn items faster and recall them longer when they logically relate to each other. Perhaps these relations allow us better to excite the appropriate circuits

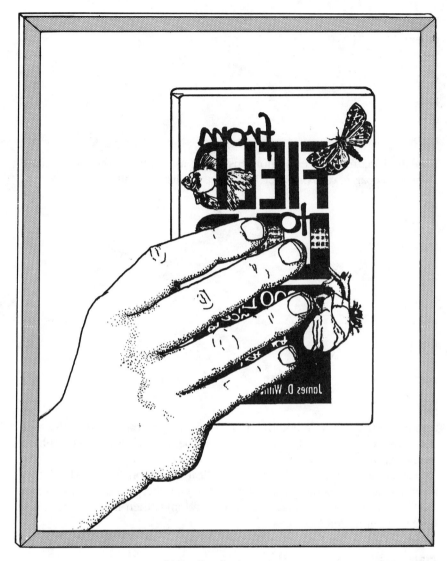

18-2   Reading in a mirror

of nerve cells. In studying, therefore, we need to connect the facts. To show how this works, copy the following words in a vertical column on an index card or lined paper:

you

as

to

you

them

others

do

would

to

do

have

Have a subject memorize this nonsensical list as you time how long it takes. Now on a separate card or paper, rearrange the list vertically as follows:

do

to

others

as

you

would

have

them

do

to

you

Again have the subject memorize it. If you wish, you can measure the time taken, but intuitively we know what will happen. Which list will the subject memorize faster and recall longer?

Many people already know the biblical quotation just cited, so they require no time to memorize it. If you wish, try another, less familiar verse for this experiment. First arrange the words randomly, then in a sentence, and have your subject memorize both.

Materials

- A speech to give
- A home with several rooms

LEARNING A SPEECH

Professional speakers sometimes memorize the order of their speeches by imagining themselves giving main points in different rooms of their homes as they walk through. By following the sequence of the rooms, they use a familiar sequence of nerve cells to remember the sequence of their speeches.

If you have a speech to give—one with several issues to cover—imagine yourself presenting each issue as you walk through your home. Better yet, place the different parts of the speech in order in different rooms, and give them in the rooms as you come to them. Rehearse the speech several times.

The Greeks and Romans first used this system over 2000 years ago, calling the system "lochi," meaning rooms. Our word locution is derived from lochi.

Materials

- People to meet and remember

REMEMBERING NAMES

We recall exciting events, colorful people, and dramatic speeches better than dull ones. To remember, therefore, dramatize what you learn. To remember names, for example, associate the people you meet with bizarre images or activities. For "Mary Wood," imagine wooden logs dancing merrily around her. For "Dr. Witherspoon," imagine a physician picking up a withered spoon. The stranger your mental image the more likely you will remember it.

To strengthen your memory further, repeat the name while conversing, and note the person's physical appearance. By doing so, you alert more regions of your brain, allowing easier recall.

Materials

- Grocery list

MEMORIZING A LIST

You can extend the room system for learning speeches and the dramatic system for learning names, to the learning of lists. If you have several items on a grocery list, for example, imagine going from room to room and finding in each a dramatically displayed food. Imagine milk turned over and spilling from your refrigerator in the kitchen, watermelon smashed on the table in the dining room, carrots being used as darts by your mother in the living room, and so on.

Materials

- Human subjects
- Three different lists of two-digit numbers

LEARNING THROUGH SOUND, SIGHT, AND FEELING

Read aloud a list of 10 two-digit numbers to a subject. When finished, ask him or her to say the numbers remembered. Record the correct answers.

Next have the subject silently read a different list of 10 two-digit numbers, reading it only once, then say the numbers remembered. Record again.

Finally read aloud a third set of two-digit numbers, having the subject write the numbers on paper as you read. When finished, remove the subject's list and ask him or her to say the numbers remembered. Record again.

How does your subject learn best, by hearing or seeing? Would other subjects learn differently? Try them to find out. Does the combination of hearing, seeing, and feeling reinforce learning? The combination engages more nerve cells than does hearing or seeing alone.

We feel writing and other movements with stretch receptors in our muscles and tendons, a process called *kinesthesis*. Would you expect note taking to improve the retention of persons who listen to lectures?

Ask the subject which numbers he or she remembers the next day. By what method does the subject retain the most numbers overnight?

Materials

- Human subject
- Pencil or pen
- Index cards or paper
- Watch with a second hand

LEARNING WITH AND WITHOUT DISTRACTION

We learn faster when we are free from distraction. You might have noticed how a phone call or an unexpected visitor disrupts your chain of thought, even when the call or visitor is for another member of the family. Perhaps the use of some nerve circuits interferes with the use of others. To show this effect quantitatively, print the following verses of Edgar Guest[1] on two separate cards, A and B.

A
Somebody said that it couldn't be done,
But he with a chuckle replied
That "maybe it couldn't," but he would be one
Who wouldn't say so till he'd tried.
B
So he buckled right in with the trace of a grin
On his face. If he worried he hid it.
He started to sing as he tackled the thing
That couldn't be done, and he did it.

Place your subject in a quiet room, perhaps your living room. Give the subject section A or B to memorize, and time how long it takes. Then give the subject the other section to memorize but with distraction. Open the door, turn on the television, turn the volume up and down, throw paper wads, and so on. What are the results?

OTHER ACTIVITIES

Some students say that they study better with music than without, and that certain kinds of music work better than others. You can test this by giving several people some poetry or prose to memorize while listening to various kinds of music, both vocal and instrumental, and during silence. What are your results?

Additionally, you can train one or more animals to negotiate a simple maze. Dogs respond simply to a pat on the head, but rodents and most other animals need greater rewards. If you want a rat to run through a maze, construct a maze with just one or two turns in it. Withdraw food from the rat one day before the test. When ready, put food at one end of the maze, and the rat at the other. It will explore the passageways until it reaches the food. Allow the animal to eat its fill, then return it to its original cage. Repeat the test daily, feeding your rat only during the tests. Record the time and number of incorrect turns during each of several runs. Eventually the rat will make no errors. It has established the correct connections between nerve cells.

Endnotes

1. From *The Path to Home* by Edgar Guest @ 1919. Contemporary Books. Reprinted by permission.

Part 6

Senses

Chapter **19**

Feeling, tasting, and smelling

*T*he color of rainbows, the warmth of campfires, the sound of music—these are life's sensations. Humans build machines that observe, record, and respond, but not with emotion. The senses and the mind place humans over matter.

Sensations differ, but the nerve impulses that transmit them are the same. All impulses are weak electrical signals traveling along nerve cells. Being alike, the impulses carry identical information about all sensations. Yet we can tell one sense from another by the type of *receptors* (sense organs) stimulated, and the part of the brain receiving the impulses. The light receptors of the eyes, for example, differ from the sound receptors of the ears, and so do the parts of the brain to which they connect. If we could cross the nerves for vision with those for sound—directing them to the wrong parts of the brain—we would see thunder and hear lightning.

What senses are there? Most people think of the classic five—sight, sound, taste, touch, and smell—but there are many more. What about warmth, cold, pressure, and pain? What about hunger, thirst, and the sense of balance? We measure balance by stretch receptors in muscles, tendons, joints, and the inner ears. What about the senses that few know about, those that measure blood glucose, oxygen, carbon dioxide, and other chemicals?

Some animals have receptors that are ultrasensitive or adapted for special purposes. Moths, for example, smell mates at distances up to 10 kilometers (6 miles). Bats emit sounds that bounce off objects and back to their sensitive ears, letting them navigate in total darkness. And blood-sucking insects locate prey from the heat of the victim's body. These are but a few of the unusual adaptations of sense organs.

Materials

- Human subject
- Red and blue felt-tipped pens with water-soluble ink
- A ruler scaled in millimeters

LOCALIZING TOUCH

How accurately can a person locate touch on the skin? Some parts have more touch receptors than others, allowing greater accuracy.

Have a subject close his or her eyes and keep them closed. Touch the palm of the subject's hand with a red felt-tipped pen. Then have the subject touch the palm with a blue felt-tipped pen as nearly as possible to the site you touched. Measure the distance in millimeters between the red and blue spots.

Repeat this procedure of touching, retouching, and measuring on several different regions of the body, such as the fingertip, face, hairless parts of the forearm, back of the neck, and sole of the foot. Where is the subject best able to localize touch?

Materials

- Sharp-pointed scissors
- A ruler scaled in millimeters

TWO-POINT DISCRIMINATION

Because some parts of the skin have a greater density of touch receptors, these parts can better discriminate between two points that you touch.

Obtain scissors with sharp points. Have your subject shut his or her eyes. Begin the discrimination test by drawing the points of the scissors together and placing them lightly on the surface of the subject's palm. Repeat with the points drawn progressively farther apart (FIG. 19-1). Determine the minimum distance between scissor points at which both can be distinctly felt.

Try the two-point discrimination test also on the fingertip, face, hairless parts of the forearm, back of the neck, and sole of the foot. The points will likely be closest together on the fingertip, where there are many touch receptors. The points will be farthest apart on the back of the neck, where there are relatively few touch receptors. If two points on the neck are closer than 25 millimeters, they are often felt as one.

Compare the results of two-point discrimination with the results of the previous experiment on localizing touch. Are they similar?

TOUCH RECEPTORS OF HAIRS

Run your fingers lightly through the hair of your arm. A special receptor for touch is located at the base of each hair. The hair acts as a lever. When bent, it

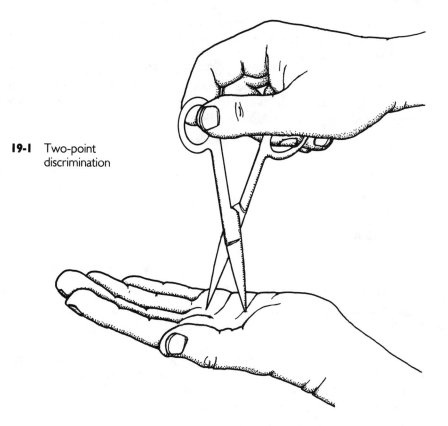

19-1 Two-point discrimination

presses upon the receptor, magnifying the touch sensation. Being longer, cats' "whiskers" magnify the sensation more than human hair, making them more sensitive yet.

Materials
• Pencil and paper

STRETCH RECEPTORS

Stretch receptors (proprioceptors) in muscles, tendons, joints, and the inner ear provide information about body positions and movements. To show this function, mark an x on a piece of paper. Raise your hand above your head, close your eyes, bring down your hand, and make a dot as near as possible to the x. Open your eyes to see how accurate you were. Raise your hand again, close your eyes, and make another dot, attempting to get closer to the x. Do this several times. You are using the proprioceptors in muscles, tendons, and joints to judge the position of your pencil. Repeat this test with your eyes open. You do better by using both proprioception and vision, especially if you make slow movements of the hand as you come near the x.

Now shut your eyes, raise your hand over your head, quickly touch your nose with a finger, and quickly touch each of the five fingers on your opposite hand. How accurate are you?

As a final test, write the word "proprioception" on a line of paper. Then place your pencil on the same line directly to the right, close your eyes, and again write "proprioception." Which word is better written? Does handwriting depend mainly on proprioception or vision?

Materials

- Human subject
- Scissors
- String
- Cotton balls
- Other small objects

IDENTIFYING OBJECTS BY FEEL

Ask your subject to close both eyes and to identify objects that you put in his or her hand. Try scissors, string, cotton balls, and so on. For a surprise, try ice and water.

The subject recognizes most of these objects by touch and proprioception. Specific regions of the brain detect and analyze the sensations. If the regions for touch and proprioception are damaged or if the sensory nerves leading to the brain are damaged, the subject will perform poorly.

Materials

- Paper towel
- Sugar

DRY AND WET TASTES

Wipe your tongue dry. Place a few crystals of sugar on it, near the tip. Is there any taste? Close your mouth and dissolve the crystals with saliva. Is there now taste? The receptors are located in miniature depressions of the tongue. Foods must dissolve to reach and penetrate these receptors.

Materials

- Human subject
- Small pieces of raw potato, green apple, beet, and turnip

SMELL AIDS TASTE

Have a subject close both eyes and pinch shut his or her nose. Place a small piece of either potato, green apple, beet, or turnip in the subject's mouth. Have the subject chew the food once or twice, note the presence or absence of taste, then identify it. Repeat the test several times, using each of the foods.

The smell of food helps identify it. Little or no taste is present when odors are absent. You might have noticed that taste disappears when you have excessive mucus in your nose during a common cold. The mucus interferes with both smelling and tasting.

Materials

- Human subject
- Thermometer
- Small beakers of water at 31°, 33°, and 35° C (87.8°, 91.4°, and 95.0° F)

COLD HANDS, WARM HEART

Heat water in small beakers to 31°, 33°, and 35° C. Touch these beakers in various orders to someone's hand and abdomen. Record the actual temperatures in the beakers, and have the subject tell which feel warm or cool in each location.

Trunk temperatures are several degrees warmer than those of the hands and feet. Thus objects that feel warm in the hand will usually feel cool on the abdomen.

Materials

- Shallow pan of ice water

REFERRED PAIN

When we feel pain, it is usually at or near the site of injury. If there is a splinter in the hand, for example, we feel it in the hand, not the foot. But in some locations, pain is referred to other parts of the body. If the arteries of the heart become blocked by lipids or blood clots, for example, the patient often feels pain in the left shoulder and arm. The pain is *referred*. Also, if an arm or leg is amputated, the patient has phantom feelings from the missing limb. Sensations in the trunk are referred to a limb that no longer exists.

Why do people have referred pain? Probably because the parts of the spinal cord and brain that receive pain-generating impulses from one region of the body also receive impulses from other regions. The impulses share a common path. Nerve cells from the heart and left shoulder, for example, transfer their information into other nerve cells that serve both the heart and shoulder. If there is pain in the heart, therefore, it might be mistakenly interpreted as coming from the shoulder. Similarly, the nerves of the leg pass into the trunk. If the leg is amputated, therefore, sensations from the severed nerve in the trunk might still seem to come from the leg.

To feel referred pain, place your elbow in a shallow pan of ice water. Be sure there is enough ice to keep the water near 0° C (32° F). Leave your elbow in the water for two or three minutes. Notice the sensations of cold and then pain in the elbow and the eventual extension of pain along the side of the arm to the fourth and fifth fingers.

Sensory branches of the ulnar nerve come from the fourth and fifth fingers through the forearm to the elbow and the upper arm to the spinal cord. Pain in the elbow, therefore, might be felt in the fingers and arm.

Materials

- Whole cloves or oil of cloves
- Other spices
- Coin
- 2-liter or larger pan of water at approximately 45° C (113° F)
- Thermometer

SENSORY ADAPTATION

We do not readily adapt to pain. To prove this, pinch your neck. It will continue to hurt as long as you keep pinching. The pain receptors in the neck continue to discharge impulses to pain-detecting regions of the brain.

In contrast, we do adapt to most other sensations. You might have noticed, for example, that when you first dive into a swimming pool, the water seems colder than it does later. The receptors for cold and most other sensations generate more nerve impulses immediately upon exposure to the sensation and fewer impulses later. You adapt.

To show that you adapt quickly to most sensations, try the following experiments:

1. Smell whole cloves or oil of cloves from the spice shelf in your kitchen. In doing this, place your nose near the source of the odor, inhale through your nose, and exhale through your mouth. Continue inhaling and exhaling in this way for one or two minutes. Does the odor become weaker as time passes? Then try the same experiment with other spices.

2. Place a coin on the palm of your hand. How quickly does the sensation of touch and pressure weaken? You do a similar experiment daily when you put on clothes. How would clothes feel if your receptors for touch and pressure did not adapt?

3. Heat a 2-liter (2-quart) or larger pan of water to approximately 45° C. Immerse your right hand in the pan, noting the sensation of warmth. After one minute, immerse your left hand in the pan while continuing to immerse the right hand. Which of the two hands now feels warmer? In which hand have the receptors for heat already adapted?

Table 19-1 Locations and Functions of Cranial Nerves

Names	Sites innervated	Actions
Olfactory nerves	Olfactory receptors in the nasal cavities	Smelling
Optic nerves	Photoreceptors in the retinas of the eyes	Seeing
Oculomotor nerves	Muscles and proprioceptors of the eyeballs	Moving the eyeballs and detecting these movements, constricting the pupils, and focusing the eyes
Trochlear nerves	Muscles and proprioceptors of the eyeballs	Moving the eyeballs and detecting these movements
Trigeminal nerves	Sensory receptors in the face and muscles of the jaws	Feeling touch, temperature, and pain in the face; chewing and detecting the movements of chewing
Abducens nerves	Muscles and proprioceptors of the eyeballs	Moving the eyeballs and detecting these movements
Facial nerves	Muscles of facial expression, salivary glands, proprioceptors, and taste receptors	Moving the muscles of facial expression and detecting these movements; salivating and tasting
Vestibulocochlear nerves	Inner ears	Hearing and sensing positions of the body
Glossopharyngeal nerves	Tongue, pharynx (throat), and salivary glands	Tasting, touching, and feeling pressure and pain in the tongue and throat; swallowing, gagging, and salivating
Vagus nerves	Respiratory tract, digestive tract, heart, and other organs of the chest and abdomen	Swallowing, speaking, and regulating heartbeat and digestion
Accessory nerves	Muscles and proprioceptors of the pharynx (throat), larynx (voice box), and neck	Moving the pharynx, larynx, and neck and detecting these movements
Hypoglossal nerves	Muscles and proprioceptors of the tongue	Moving the tongue and detecting these movements

OTHER ACTIVITIES

Twelve pairs of cranial nerves emerge from the underside of the brain. Each nerve is formed by thousands of nerve cells. Some of the nerves are *sensory*: They carry impulses from sensory receptors to the brain for analysis. Others of the nerves are *motor*: They carry impulses from the brain to the muscles and glands, causing the muscles to contract and the glands to secrete. Still other nerves are *mixed*: They carry some impulses to the brain, others to the muscles and glands.

TABLE 19-1 tells what the cranial nerves do. To find whether the nerves and brain are working, physicians sometimes ask their patients to respond to specific sensations or to contract specific muscles. For example, an ophthalmologist might ask a patient to read an eye chart and visually to follow the movements of a finger. If the patient can do so, the photoreceptors, sensory nerves, visual parts of the brain, motor nerves, and muscles of the eyeballs are all working.

Look at the table of cranial nerve locations and functions. Devise tests, similar to that just described, to show whether each cranial nerve is working, and try them.

Chapter **20**

Anatomy
of an eye

"Who could believe," said Leonardo da Vinci, "that so small a space could contain the images of all the universe? O mighty process!"[1] Da Vinci (who lived in the 1500s) was describing the eye in which René Descartes later saw the images on the retina. To see them, Descartes scraped away the *sclera*, the outer covering, from an ox eye, allowing him to peer into the *retina*, the light-receiving, inner surface (FIG. 20-1). Then he mounted the eye in the knothole of a wooden shutter. Looking into the retina, he saw an inverted image of the outdoors.

The eye, therefore, acts like a camera. Both use a lens to focus an inverted picture on the surface behind the lens. In the eye the surface is a retina; in the camera, it is the film.

Willy Kühne, in the 1800s, compared the pigments in the retina to the emulsion of the film. He studied *rhodopsin*. This is a purple pigment that bleaches to yellow when exposed to light, a process, he assumed, that made the retinal image "seen" by the brain. In the dark, the pigment reconverts from yellow to purple. "The retina behaves not merely like a photographic plate," said Kühne, "but like an entire photographic workshop, in which the workman continually renews the plate by laying on new light-sensitive material, while simultaneously erasing the old image."[2]

The analogy of the eye to the camera goes beyond the retina and film. There is an *iris* in the eye—a colored, muscular ring around the pupil—which acts like an iris diaphragm in a camera. Both irises control the amount of light entering, opening more in dim light and less in bright light. Also, there is a black surface in both the eye and camera. The blackness in the eye results from *melanin*, a pigment at the back of the retina. By absorbing light, the black melanin of the eye and the black surface of the camera prevent overexposure and blurring of images.

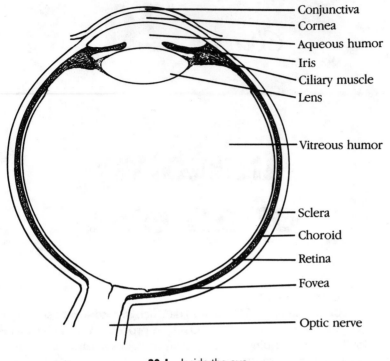

Conjunctiva
Cornea
Aqueous humor
Iris
Ciliary muscle
Lens

Vitreous humor

Sclera
Choroid
Retina
Fovea

Optic nerve

20-1   Inside the eye

The eye and brain also have features unlike those of cameras. For one, the brain "sees" nerve impulses, not retinal photographs. For another, the lens of the human eye focuses by changing its width, not by moving nearer or farther from the retina as the lens of a camera does from the film. For yet another, the retinal image is always moving.

One would think that the retinal image would be stationary when the eye looks at a stationary object. Instead, the eye quivers slightly and roams over the surface of the object to identify it, moving the image on the retina. When you look at a face, for example, your eyes shift from the eyes of the face to the nose, mouth, and so on. The movements of the eyes allow images to shift rapidly from one group of photoreceptor cells to another. This shift prevents excessive bleaching of the visual pigments in any one area. When experimenters hold the eyes still, continuous bleaching causes the image to fade and disappear.

Materials

- Mirror
- Flashlight

EXAMINING YOUR EYE

Look into a mirror to see the external features of your eye. The colored ring-shaped structure is the *iris*; the black circular opening is the *pupil*; and the

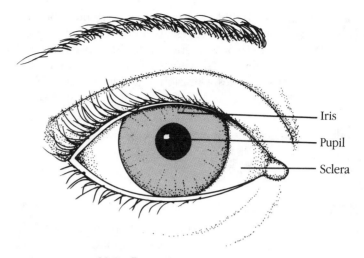

20-2   External anatomy of the eye

white of the eye is the *sclera* (FIG. 20-2). The iris contains two sets of muscle fibers—circular and radial—which regulate the size of the pupil.

As you continue to look in the mirror, shine a flashlight into your eye. The circular muscle of your iris contracts, narrowing the pupil. The constriction reduces the amount of light entering the eye. Now turn off the flashlight. The circular muscle of your iris relaxes and the radial muscle contracts, pulling the iris outward. What does this do to the pupil? Of what benefit is the admission of more light?

In watching the iris and pupil, you look through two transparent layers that cover them. The first is the thin, sheet-like *conjunctiva*. This layer covers the entire part of the eye that you see in the mirror, extending from the upper eyelid across the eye to the lower eyelid. If you place your finger under the lower eyelid, and gently pull the lid downward, you will see the site at which the conjunctiva of the eye joins the lower lid. By joining the eye to the eyelids, the conjunctiva keeps dirt and insects from getting behind the eyeball.

The second transparent layer, the *cornea*, lies under the conjunctiva and over the iris and pupil. The surface of the cornea is convex—that is, it is rounded and protrudes slightly outward—and the cornea is much thicker than the conjunctiva. When light enters the cornea, the rays are bent inward toward the transparent *lens* that lies immediately behind the pupil.

The perimeter of the cornea joins the sclera or white of the eye. The word sclera means hard. You see only part of this tough, white, fibrous membrane when looking in a mirror. The rest of it continues around the entire globe-shaped eyeball.

Notice the shiny surface of your eye. The shininess results from a thin layer of transparent fluid. This fluid comes from *lacrimal glands*, the tear-secreting glands that you cannot see. They are concealed behind the upper eyelid. Every time you blink, lacrimal fluid flushes across the conjunctiva, keeping it moist.

Notice also the tiny red blood vessels that carry nutrients and oxygen across the sclera of your eye. These vessels do not penetrate the transparent cornea or lens where they would interfere with the transmission of light. Yet the cornea and lens are living. Their transparent cells consume oxygen and nutrients and dispose of wastes. Diffusion to and from distant blood vessels keeps them alive.

Materials
- Owl or other bird
- Sheep or other grazing animal

NICTITATING MEMBRANE

Many terrestrial animals have a third eyelid, a transparent eyelid that moves sideways across the eye. This lid is the *nictitating membrane*. Birds close it during flight, protecting their eyes while allowing them to see. Grazing animals close it to prevent abrasion of the cornea by rough vegetation. Humans do not have a third eyelid.

Look for the transparent, nictitating membrane in owls or other birds and in sheep or other grazing animals. If such animals are not available, look for the membrane in a dog as it sleeps with its other eyelids partially open. The nictitating membrane closes from the nasal side to the lateral side of the eye.

Materials
- Forceps and sharp-pointed scissors
- Dissecting tray
- Surgical gloves
- Eyeball of a sheep or other mammal

DISSECTING AN EYEBALL

Obtain either a fresh or preserved eyeball of a sheep or other mammal. If you want a fresh eye, contact a custom butcher or meat packer listed in the yellow pages of your phone book. If you want a preserved eye, contact one of the biological suppliers of Appendix B.

CAUTION: Handle the preserved eye with surgical gloves. Use water to thoroughly rinse it, and place it in a dissecting tray. If you get preservative on your skin, wash it off several times with soap and running water. Then keep your hands away from your eyes.

Look first for the cornea and sclera (FIG. 20-1). If the eye is fresh, the cornea will be transparent; if it is preserved, the cornea will be nearly opaque. In either case, the adjacent sclera will be white, tough, and continuous with the cornea.

Look through the cornea to see the pigmented ring-shaped iris and the pupil it surrounds. The anatomy closely resembles that of humans.

Several strap-like, red, cream, or brown muscles attach externally to the sclera. In life, these muscles move the eyeball up, down, obliquely, or sideways. Fatty tissue lies between the muscles, obscuring them. Tease and cut away the fat for a clearer view.

Look behind the eye at the side opposite the cornea. You will see the short, white stub of an *optic nerve* exiting the sclera. The optic nerve transmits signals from the light receptors of the retina to the brain.

Using sharp-pointed scissors, cut through the thick-walled sclera and around the eyeball, separating the eye into two parts: an upper third containing the cornea, iris, and usually the lens, and a lower two-thirds containing the other structures (FIG. 20-3). Sometimes the lens is drawn into the lower two-thirds. When opening the eye, cut carefully to prevent the fluid of the eye from squirting on you. The pressure of the fluid causes the eye to be globe-shaped instead of flat and, in life, keeps the retina smooth. Wrinkled retinas distort vision.

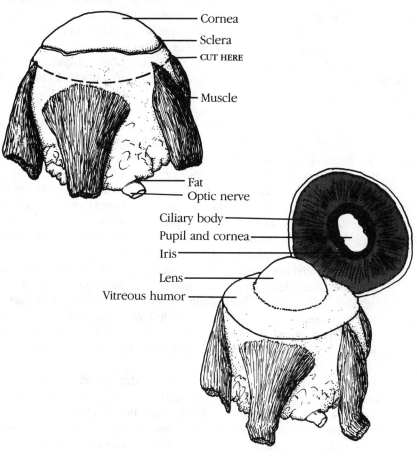

20-3 Cut around the eyeball

Beginning with the upper third of the eye, observe first the convex, almost spherical lens. Leave it attached for now.

Light rays bend as they pass through the living cornea, and bend again as they pass through the lens. The bending focuses the rays on photoreceptors of the retina at the back of the eye. This effect resembles that of a magnifying glass. Both lenses focus light.

If you are studying a preserved eye, its lens will be opaque, as in the cornea. *Cataracts* of living lenses also cause opacities. Such cataracts develop occasionally in children but more often in old people. The lens continues to grow slowly with age, and as it grows the cells at the center become further removed from the surface. Because oxygen and nutrients diffuse through the surface to reach the center, and because wastes diffuse from the center to reach the surface, growth of the lens lengthens the diffusion pathway. This pathway becomes so long that wastes collect and protein degenerates in the central cells, causing these cells to become opaque. The opacity makes it difficult to see.

The lens itself is surrounded by a thickened black ring of radially oriented ridges called the *ciliary body*. This body connects through extremely short, delicate, threadlike fibers to the lens. The fibers hold the lens in place.

A ring of *ciliary muscle* lies inside the ciliary body that encircles the lens. You probably cannot distinguish this muscle from the black substance around it. When the ciliary muscle contracts, it squeezes the threadlike fibers inward, allowing the lens to become more convex, more nearly spherical. The nearly spherical lens focuses light from nearby objects.

Gently tear or cut loose the black ciliary body from the lens. Notice the opening of the pupil in front of the lens and the dark iris surrounding the pupil. As you remove the lens, a watery fluid, the *aqueous humor*, spills out. This fluid is in the cavities between the lens, iris, and cornea.

In life, the ciliary processes secrete the aqueous humor, and thin-walled veins drain it away. If too much fluid is secreted or too little drained, a person develops *glaucoma*. In glaucoma, the fluid pressure increases so much that it presses against arteries lining the inside of the eye, hindering blood flow. If the flow is sufficiently reduced, the victim becomes blind.

Turn now to the other portion of the eye, the posterior two-thirds that you cut away earlier. It is filled with a gelatin-like mass called the *vitreous humor*. Gently remove this humor. Under it, along the concave surface of the eye, is the smooth or crumpled, cream-colored layer called the *retina*. In life, the retina adheres to the pigmented *choroid* layer that lies beneath it. But in death and dissection, the retina usually breaks free from the choroid, except where the retina attaches to the optic nerve.

The retina contains the *photoreceptor cells* that detect visual images. Sometimes these receptors are called rods and cones because of their shapes. Light is focused on the receptors. Their pigment bleaches in response, generating electrical impulses that pass through the optic nerve to the visual parts of the brain. The brain interprets the impulses as visual images.

The choroid layer, located between the retina and sclera, is entirely black in humans but partly iridescent in sheep. Iridescence occurs more often in animals active at night than active by day. To see the effect of this, shine a flashlight toward the eyes of a cat at night. Its iridescent choroid reflects the light. The reflection exposes the retina twice to the same light rays, producing a stronger image. In other words, the reflective choroid makes the eyes more sensitive in dim light.

If you want to keep the eye, put it in a jar of isopropyl alcohol. Otherwise, discard it and your surgical gloves in a trash container. Wash your hands thoroughly with soap and water.

OTHER ACTIVITIES

Among animals, different kinds of eyes serve different purposes. Use a hand lens or microscope to examine various eyes, including the compound eyes of insects.

Each eye of an insect has hundreds or thousands of separate, tightly packed, tubular units called *ommatidia*. At the surface of the eye, the ommatidia resemble hexagonal tiles in a floor. Each ommatidium has its own lens system to concentrate light, its own pigment cells to prevent the scattering of light, and its own photoreceptors to convert light into nerve impulses. Because an eye contains hundreds or thousands of separate ommatidia, the insect readily perceives movement—in which the image crosses and penetrates the separate visual units—but does not see the total image clearly. Instead it sees a visual mosaic.

Study also the placement of eyes on the heads of different animals. Notice, for example, that human eyes are closely placed at the front of our heads. This location provides an overlap of visual fields, making it easier for us to work with three-dimensional tools. Similarly, the overlapping visual fields of monkeys allow them to judge the three-dimensional positions of tree limbs. In contrast, the eyes of horses and rabbits are distantly placed at the sides of their heads. This allows them to see predators from the front, side, or back.

Endnotes

1. Da Vinci, Leonardo. Vol. 1, Ch. 9 of his notebooks. Translated by Edward McCurdy.

2. Kühne, Willy. Quoted by George Wald. "Eye and camera," *Scientific American*, 185 (August 1950): 37-38.

Chapter **21**

Vision

*H*ave you ever poked a stick at an angle into water? If so, have you noticed that it seems to bend where it meets the water? It seems bent because light rays change directions when passing from one transparent medium to another.

Light also bends when passing from air into the *cornea*, the transparent cover of the eye (FIG. 21-1). It bends again when passing from the fluid of the eye into the *lens*. The bending of light focuses the visual image on the *retina*, the light-receptive lining of the eye.

To focus from near to far objects, we change the shape of the lens. To see near objects, we make it almost spherical; to see far objects, we make it flatter. Children have elastic, flexible lenses, allowing them to focus clearly on objects both near and far. Adults have less elastic lenses, making it difficult for them to focus nearby. This is why older people hold books far from their eyes.

Human eyes work best in moderately bright light. Researchers showed this by measuring muscle tension and eye blinks in people exposed to different illumination. In one experiment, subjects read print illuminated by 10 footcandles (common home lighting) and later by 100 footcandles (shade on a sunny day). In the brighter light, there was 11 percent less muscular tension and 12 percent less blinking. In another experiment, an industrial plant increased general illumination from 2 to 20 footcandles. Accidents then decreased 50 percent.

These and other experiments show that we often need:

1. *More light*. We should use the most powerful fluorescent and diffused incandescent lamps that are commonly available.

2. *Larger print*. We should choose newspapers, magazines, and books that have print large enough to read comfortably.

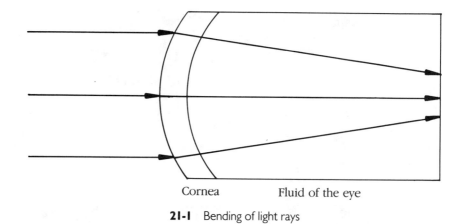

Cornea   Fluid of the eye

21-1 Bending of light rays

3. *Reduced glare*. We should face away from bright lamps and avoid reflections from television and computer screens.

Materials

- Pencil
- Bowl of water
- Hand lens

BENDING OF LIGHT RAYS

Immerse a pencil at an angle into a bowl of water. View the pencil from the side. Notice that it seemingly bends where it meets the water. This is the site where light rays actually do bend—that is, change direction—when passing between the air and water.

Use a hand lens to focus the sun's rays to a point on the ground. The rays bend in passing at an angle from the air into the curved surface of the glass. The greater the curve, the greater the bend.

Light rays bend in passing from any transparent medium into another of different density. Thus the light passing from the air into the cornea and lens is bent onto the retina.

Materials

- Figure 21-2 or Snellen eye chart

TESTING YOUR VISION

Use FIG. 21-2 or a Snellen eye chart to test your visual acuity and that of others. The Snellen chart is the one with the big letters at the top and progressively smaller letters toward the bottom. Place FIG. 21-2 or the chart 20 feet away in good light. If your vision is normal you can read the row of letters marked 20/20

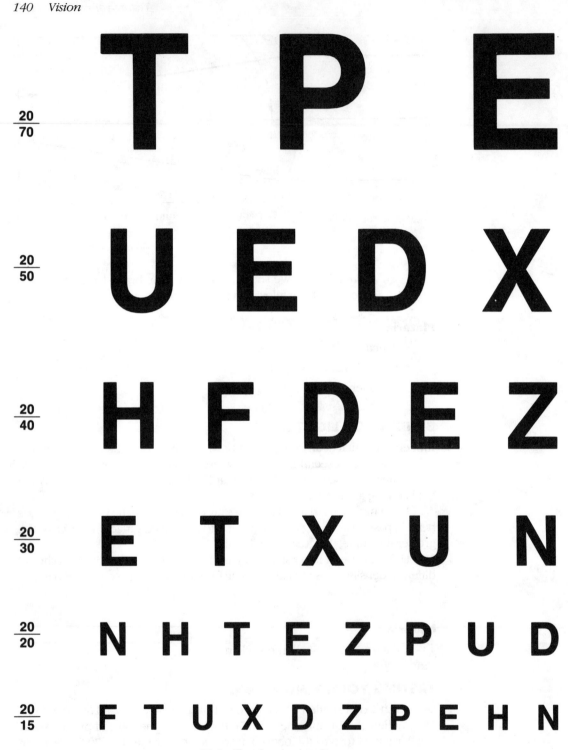

21-2 Test for visual acuity

at 20 feet. Your vision is said to be 20/20. If your distant vision is poor, you might read only the row of letters marked, let us say, 20/40 or 20/50 feet, though you are standing at 20 feet. In this case, your vision is 20/40 or 20/50. Test the two eyes together, then separately by covering one and then the other with your hand. If you wear eyeglasses or contact lenses, test your eyes with and without them.

People who have *myopia* (nearsightedness) cannot read the small letters of FIG. 21-2. They see what is near but not what is far, because light focuses in front of their retinas. Their eyeballs are too long or their focusing mechanism too powerful. To correct this problem, they wear eyeglasses or contacts that diverge the light rays before they get to the natural lenses.

Most states require 20/40 or better vision for driving. Drivers who have poorer vision must wear corrective lenses.

Materials

- Ruler, yardstick, or meterstick

HOW NEAR CAN YOU SEE?

Hold this book farther than usual from your eyes. Focus on one of its letters. Then move the print slowly toward your eyes until the chosen letter first becomes blurred. Measure this distance and compare it with the averages for other people (TABLE 21-1). Is your near point about the same as that of others your age?

Table 21-1 Near Points at Different Ages

Age	Centimeters	Inches
10	7.5	2.9
20	9.0	3.5
30	11.5	4.5
40	17.2	6.8
50	52.5	20.7
60	83.3	32.8

Test others, preferably to include family members of greatly different ages. Does the age of these members affect their near points?

People who have *hyperopia* (farsightedness) cannot read nearby print. They see what is far but not what is near, because light focuses behind their retinas. Their eyeballs are too short or their focusing mechanism too weak. Older people, for example, have rigid lenses that they cannot focus. To correct such problems, people wear eyeglasses or contacts that converge light rays before they get to the natural lenses.

Materials

- Figure 21-3

TESTING FOR ASTIGMATISM

First close one eye, then the other while looking at FIG. 21-3. Hold the illustration at arm's length from your eyes. Are all the lines clear? To a person with astigmatism, the lines in one or more planes will be blurred and not as dark as the others.

21-3   Test for astigmatism

Astigmatism is a condition of blurred vision resulting from an unequal curvature of the cornea or lens. It is corrected by specially ground lenses.

Materials

- Sheet of paper

TWO VIEWS FOR THREE DIMENSIONS

Humans, cats, and many predators have both eyes at the front of the head rather than the sides. This position allows the visual fields of the two eyes to overlap. When the visual data passes from the two eyes to the brain, the brain translates the two images into a single image having three dimensions. To show that the images seen by the two eyes are in different positions, look at some distant object. Form a circle (okay sign) with your index finger and thumb through which to view this object. Hold the circle 10 to 30 centimeters (4 to 12 inches) from one of your eyes. Close first one eye and then the other. The circle seemingly moves to and from the object, proving that you see different views. Only one of the two eyes sees the object through the circle.

Because the eyes see different views, you can make your hand appear to have a hole in it. Roll a sheet of paper into a tube of 3 to 4 centimeters diameter (about 1.5 inches). Keep both eyes open. Hold the paper tube in your left hand directly in front of your left eye. Open your right hand. With the right palm facing you, place the side of the hand against the surface of the tube. Does the right palm now appear to have a hole in it? If not, move your palm backward and forward along the tube.

Materials

- Human subject
- Needle
- Thread

NEED FOR THREE-DIMENSIONAL VISION

To show the convenience of three-dimensional vision, thread a needle while keeping both of your eyes open. Next thread it with your right eye closed, then your left. In each case, how many attempts does it take to thread the needle? With one eye closed, the view resembles that on an ordinary, two-dimensional television or movie screen.

Ask another person to face you. Extend the index fingers of your right hands toward each other until the fingertips touch. With your eyes open and three-dimensional vision intact, this should be easy. Now each of you close one eye and again touch your fingertips. What happens?

Materials

- Human subject
- Pencil or index finger

CONVERGENCE OF THE EYES

Our eyes converge—rotate toward the nose—when focusing on near objects. As they converge, our brains continue to interpret the two views of each object as a single image until we exceed the limits of convergence.

To show convergence, hold a pencil or an index finger upward at arm's length, and focus on its tip. Now move it slowly toward your nose until a double image appears. Repeat this test on another person as you watch his or her eyes. Do they converge?

Materials

- 5 to 10 sheets of white paper
- Rubber band
- Mirror

ADAPTATION TO DIM LIGHT

In dim light, our pupils dilate (get bigger) and the light-receptor pigments of our retinas regenerate. These adaptations allow more light to enter the eyes and make the retinas more sensitive to the light that does enter.

Roll 5 to 10 sheets of white paper lengthwise into a single hollow tube in which all sheets extend twice around the tube. Slip a rubber band around the tube to hold them in place. Press the bottom of the tube firmly against the print of this page, and one of your eyes firmly against the top of the tube, permitting little or no light to enter the tube from either end. Place your free hand over the other eye. The light should be so dim in the tube that you cannot read the words but can see that they are present. If there is too much or too little light, add or subtract a few sheets of paper. Continue to look down the tube at the words for two or three minutes. Your eye gradually adapts to the dim light, making the words become clear enough to read.

After you read the words, look immediately at a mirror. Your pupils dilated as you peered into the dim tube and as you covered the free eye. Now that the pupils are exposed to bright light, they constrict (get smaller).

Have you noticed how dark a movie theater appears when you first enter it? If the movie has started, do you have difficulty seeing the seats? After a few minutes, does the theater appear brighter? Your pupils and light receptors have adapted.

Materials

- Stars in the nighttime sky

HOW BEST TO SEE FAINT STARS

The retina of the eye contains two kinds of light receptors—*rods* and *cones*—named for their shapes. Rods are sensitive to dim light, cones to bright light.

The rods are concentrated at the periphery of the retina and the cones near the center. Indeed, when we look at any point, the light from that point is focused directly on the *fovea*, a depression in the retina that contains only cones. Because the fovea has only cones, it cannot see dim light. Thus, if we look directly at a point of dim light, we cannot see it.

On a dark night, look directly at a faint star, one that is barely visible. Your lens focuses the starlight on your fovea—where there are only cones and no rods—so the star disappears. Now look slightly to one side of the star. It reappears! Its light now falls on rods that lie outside the fovea.

Materials

- Figure 21-4

BLIND SPOT

The rods and cones of each eye connect to nerve cells that pass from the retina through an optic nerve. There are no rods and cones—no way to see—at the place where the optic nerve leaves the eyeball.

Because there are two optic nerves, one for each eye, we each have two *blind spots*—two places that cannot see. We do not notice the blindness because each of our two eyes sees slightly different visual fields. The gaps are covered.

To identify your blind spots, use the cross and face diagram of FIG. 21-4. Close your right eye while looking at the cross with your left eye. At about 30 to 40 centimeters (12 to 15 inches) from your eye, the face will disappear. Move the page back and forth to find the exact distance. Now close your left eye while looking at the face with your right eye. You have found the blind spots at the back of the retina where there are no receptors for light.

21-4   Test for blind spot

Materials

- Frosted 40-watt light bulb
- White, red, green, blue, and yellow paper
- White poster board or other white surface
- Green, black, and yellow crayons

AFTERIMAGES

When we stare at a bright object, then look away, the light receptors of the retina continue to discharge. This discharge produces a bright *afterimage*. As the discharge continues, the pigments of the photoreceptors become bleached, causing the image to darken.

Turn on a frosted 40-watt light bulb. Gaze steadily at the light for three seconds, then glance away, preferably at a white surface. Does the image persist? Does it change color as time passes?

There are three types of light-receptive cones in the human retina. One type responds best to red, another to green, and another to blue. When we stare for many seconds at a single color, such as red, the pigment in the color-sensitive cones becomes bleached. If we then look at a white surface, the complimentary color appears in the afterimage. If we look at red, we later see green.

Stare for one minute directly at a small piece of red paper or other red object. Then look at a white background, such as a white poster board. Do you see an afterimage? What color is it? Repeat this procedure for small pieces of green, blue, and yellow paper. What colors are the afterimages?

Draw an American flag on a sheet of white paper, but color just one black star on a yellow background in the upper left corner of the paper. Then color alternating stripes of green and black at the right and bottom of the paper. Gaze intently at this atrocious-looking flag for one minute, then look at a white background. A red, white, and blue flag appears in all its old glory!

Materials

- Figure 21-5
- Ruler

VISUAL ILLUSIONS

What we see depends partly on the background with which we compare it. The moon, for example, seems much larger when we see it near the horizon, not overhead.

Examine the illusions of FIG. 21-5. Are the horizontal lines of the first two illustrations of equal or unequal length? Measure them to see. Do the horizontal lines of the third illustration bend? Place a straight edge along them to find out.

Materials

- Sleeping dog, cat, or person

EYE MOVEMENTS

Each eyeball has six muscles that control its movements—up, down, sideways, or obliquely. To see horizontal eye movements, watch someone read. To see vertical and oblique eye movements, watch someone seated near a basket at a basketball game.

When we dream, our eyes follow the activity of the dream. When we sleep without dreaming, our eyes are still. To see the eye movements of dreaming, look at the eyelids of a person or pet that is sleeping. In humans, dreams occur about every 90 minutes. In dogs, the eye movements are often accompanied by limb movements, letting you know when to look.

OTHER ACTIVITIES

When you next have an eye examination, ask questions. Why does the doctor use a drug to dilate your pupils? What is the purpose of the glass lenses with which you are tested? Why puff air in your eyes? Why are you told to follow the movements of the examiner's finger?

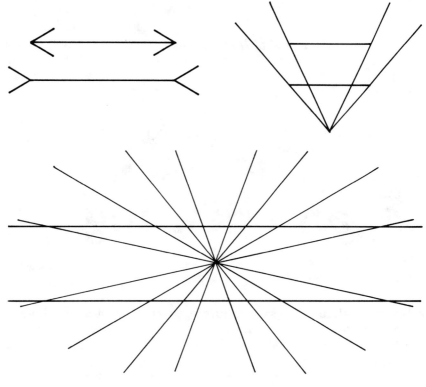

21-5   Visual illusions

Perhaps the doctor or an assistant will show you how to use an *ophthalmoscope*. This is a hand-held device that projects a beam of light into the eye, permitting the examiner to see the interior. Using it, you can see the vessels that supply blood to the retina, the blind spot where the optic nerve enters the retina, and the fovea where cones provide the sharpest vision. By examining these, the doctor finds evidence for arteriosclerosis, diabetes, cataract, and other diseases.

Chapter **22**

The ears and hearing

*B*eyond the obtrusive flaps we call ears are structures as intricate as those in fine Swiss watches—vibrating membranes, tiny bones, receptor cells, and auditory nerves—all embedded deep in the skull. With them we hear around corners, through darkness, outside the limits of the other senses.

Our ears are divided into outer, middle, and inner portions (FIG. 22-1). The outer ear consists of a flaplike *pinna*, the part most people call "the ear," and a tubular *auditory canal* through which the pinna channels sound to a *tympanic membrane* or eardrum. When sound waves reach the eardrum, they vibrate its membrane.

Three little bones—called *ossicles*—are in the middle ear. Each bone presses upon the next to transmit the vibrations of sound to the inner ear.

A small opening—the *auditory tube*—leads from the middle ear to the pharynx, the air passageway into the throat. The auditory tube opens to equalize pressure between the middle ear and the pharynx when we fly, dive, or otherwise change altitude. As it opens, we feel the release of pressure. It also provides a passage where infections spread from the nose and throat to the ear.

The third bone of the middle ear presses on a distensible membrane, the *oval window*. This window is smaller than the eardrum and moves back and forth through a greater distance. It provides the entrance to the cochlea of the inner ear.

The vibrations of the oval window penetrate the liquid-filled, snail-shaped *cochlea*, where there are receptor cells for sound. Vibration of these cells generates electrical impulses in the *cochlear nerve* to which they ultimately attach. As the impulses arrive in the brain, they are interpreted as sound.

Experimenters have placed electrodes in receptor cells of the cochlea to measure the response of these cells to different pitches. They found that cells at the base of the cochlea detect high pitches; those at the apex, low pitches. This

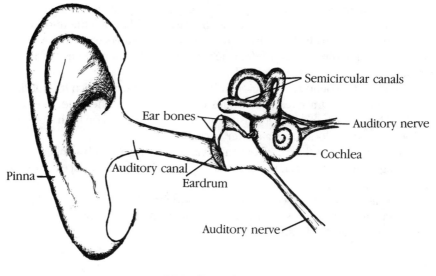

22-1 Parts of the ear

is logical because the receptor cells stand on fibers that are short at the base of the cochlea and long at the apex. The fibers resemble strings of a harp. Plucking the short strings produces high notes; plucking the long strings produces low notes.

Materials

- Tuning fork
- Bowl of water

SEEING SOUND WAVES

When sound passes through air, it generates waves. If the sound is pure—as from the tap of a tuning fork—the waves are evenly spaced. As the vibrating prongs of the fork swing forward, they push air molecules together in front of them, condensing the air. Simultaneously, they pull air molecules farther apart in back of them, rarefying the air. The alternating condensation and rarefaction vibrate the eardrum.

You cannot see air vibrating but you can see water vibrating. Tap the prongs of a tuning fork, and stick their tips in water. Watch the formation of waves as the prongs move back and forth.

Materials

- Human subject

WHERE IS THE SNAP?

Ask a subject to close both eyes. Snap your fingers at various positions around the head of this person and have him or her point each time to the source of the sound. Sometimes snap your fingers in the same spot to find whether the subject points repeatedly at this spot. Particularly do this along the midline of the head, that is, equally distant from the two ears. Most subjects guess correctly when the snap is directly to the left or right, but fail as the sound comes near the midline—in front, above, or behind.

We localize sound subconsciously by noting the side at which it is louder and at which it first arrives. Our brains do the analysis. When the sound is midway between the ears, its intensity and distance are equal on both sides. In this case, accurate location is a matter of chance.

Have the subject close both eyes again and plug one ear with a finger. Snap your fingers around the head as he or she points to the source. What effect does a plugged ear have?

The ability to locate sounds promotes survival. Both predators and prey find each other in part by sound, and we humans find it more than convenient to know the sources of traffic noise.

Materials

- Human subject
- Watch or clock that ticks

TEST FOR DEAFNESS

Gradually move a ticking watch or clock away from an ear of a subject, and record the distance at which it was last heard. Do the same for the other ear. Then have the subject test you. Is hearing equally acute in the left and right ears? Do you hear as well as your friend? Probably there are some variations.

Imitate partial deafness by plugging one of your ears with your finger. What effect does this have on the distance at which you hear the tick?

There are three categories of partial or total deafness: (1) conductive, (2) nerve, and (3) central. You experienced *conductive deafness* by plugging your ear; others experience it with aging. It results from any interference with the transmission of sound from the outer ear to the cochlea, often from the growing together of bones in the middle ear. Hearing aids—which conduct sound through bones of the skull—help this condition immensely, as does surgery. *Nerve deafness*, in contrast, results from damage to the receptor cells for sound or to the cochlear nerve. Loud noises, for example, can damage the receptors. *Central deafness* results from damage to the auditory regions of the brain, as from head injuries or strokes.

Materials

- Tuning fork

SECOND TEST FOR DEAFNESS

Strike a tuning fork, causing its prongs to hum. Immediately press the base of the fork against the midline of your forehead. You will hear vibrations transmitted both through the air and through the bones of your skull. If your hearing is normal or if it has deteriorated equally in both ears, the sound will be equally loud in both ears. If instead you are nerve deaf in one ear, the sound will be louder in the opposite ear. Or if you are conduction deaf in one ear, the sound will be louder in the same ear (because the opposite ear hears room noises as well as the tuning fork). You can simulate conductive deafness by sticking a finger in one ear to block external sound.

Materials

- Tuning fork

THIRD TEST FOR DEAFNESS

Again vibrate a tuning fork, but this time hold its base against the *mastoid process*, the bony swelling of the skull located immediately behind the earlobe (FIG. 22-2). Listen until the hum of the prongs just disappears, then hold the vibrating prongs near the auditory canal of your ear. Do you still hear the hum—most people do—or is all quiet? People with conductive deafness hear nothing. Because their ossicles have grown together or for other reasons, they cannot adequately conduct faint sounds.

As another means of testing, reverse this procedure. Vibrate the tuning fork near your auditory canal until the sound disappears. Then immediately

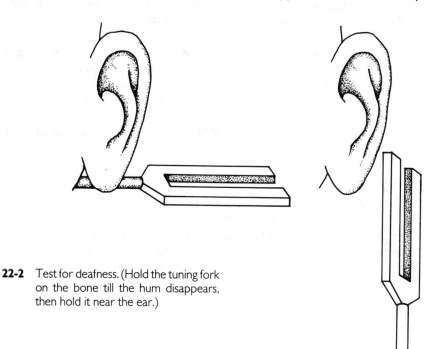

22-2 Test for deafness. (Hold the tuning fork on the bone till the hum disappears, then hold it near the ear.)

hold the base of the fork on your mastoid process. If the sound reappears, you have conductive deafness.

Do both tests on both ears.

Materials

- Flying airplane, car driving over mountains, or deep-water swimming pool

AUDITORY TUBE

The auditory tube joins the middle ear to the nasal part of the pharynx, the air passageway above the throat (FIG. 22-1). Usually the air pressure in the middle ear is about the same as that in the pharynx or outside air. But when we change altitude or depth while flying, driving, or swimming, the pressures become unequal. If the auditory tube is open, the pressurized air then moves through it, allowing the pressures to become equal.

Next time you are flying, diving, or driving up and down mountains, notice your feelings as you change altitude. As you ascend, the air or water pressure decreases in the external environment and in your pharynx. The pressure of air in your middle ear then exceeds the pressure outside, forcing air from your middle ear through the auditory tube to the pharynx. If the tube is closed when the air starts moving, you will hear a popping noise.

As you descend, the reverse occurs. Air flows from the pharynx through the auditory tube to the middle ear. Again you might hear a pop as the air moves.

If you have a cold, your auditory tube might be inflamed, making it difficult to open. In this case, notice the feeling of pressure and perhaps pain in your head, and the partial loss of hearing. If you have continued difficulty in clearing the auditory tube, chew gum or yawn. These maneuvers usually open it.

Do not purposely fly or dive when you have a cold. If the inflammation is severe, you risk rupturing the eardrum.

OTHER ACTIVITIES

Hearing clinics and sometimes university laboratories have audiometers with which to test hearing. Using these devices, examiners vary both the intensity and pitch of sounds. They test hearing over the entire scale to find the extent of deafness. Many people lose their ability to hear high pitches as they age. Whatever your age, if you have an opportunity, take an audiometer test.

Chapter **23**

How animals balance

*H*ave you heard the expression "fly by the seat of your pants?" Pilots sometimes judge their positions by the pull of gravity on their bodies. This pull activates *pressure receptors* in their seats, and *proprioceptors* (stretch receptors) in their muscles, tendons, and joints. In the early days of flying, pressure and proprioception were much used to judge position.

There is no seat from which to fly when swimming underwater; the body is floating. Pressure receptors are inactive because water supports both the body and receptors. Proprioceptors in muscles, tendons, and joints are inactive because water also supports them. Only two senses remain with which to judge position: the *eyes* and the balance mechanism of the *inner ears*. When swimming underwater with the eyes open, people can easily follow straight lines and maintain appropriate depth. With the eyes closed, however, they sometimes become disoriented. Only their inner ears maintain balance.

Deep in the skull, the inner ears contain two sensory mechanisms: one for hearing, the other for balance. The balance mechanism has three parts: a utricle, saccule, and semicircular canals (FIG. 23-1).

The *utricles* and *saccules* of the two ears are activated by changes in body position, as from lying to standing, and by linear acceleration, as from the start of elevators. Each utricle or saccule contains tiny, rocklike particles suspended at the ends of hairs in an organ called the *macula*. Changes in gravity cause the particles to bend the hairs, activating sensory receptors that direct nerve impulses to the brain. These impulses disclose body position or the direction of acceleration.

In contrast, *semicircular canals* are activated by circular instead of linear motions. When an individual turns, fluid moves through the canals, bending small hairs in the *crista*, an organ similar to the macula. The bent hairs activate sensory receptors, generating nerve impulses to the brain. Thereby the brain is told the direction of the turn.

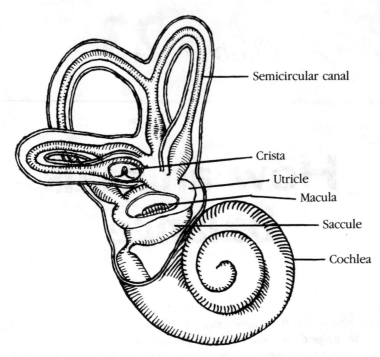

Semicircular canal

Crista

Utricle

Macula

Saccule

Cochlea

23-1   Balance mechanism of the ear

Materials

- Frog
- Aquarium or deep basin of water
- Large beaker or similar container

BALANCE IN A FROG

Animals prefer to be right side up and in their normal positions. They immediately sense and resist changes in these positions. For example, when cats inadvertently fall upside down from a tree, they turn over in midair to land right side up.

You can study this *righting reflex*, as it is called, in a frog. Begin by setting the frog on the floor to see its normal posture and hopping. Then turn the frog on its back to find how quickly it rights itself (turns over). Righting is a response to pressure on the back and to abnormal stimuli in the upside-down eyes and inner ears.

Put the frog in an aquarium or deep basin of water to see it swim. Then turn the frog on its back in the water. Does it right itself? Water exerts little or no pressure on a floating animal. With this in mind, do you think pressure receptors are essential for righting?

Move the frog to a large, empty beaker or similar container. Tilt the beaker forward, backward, and sideways. How does the frog compensate for these changes in position? What senses did it use to produce these adjustments?

Materials

- Human subject

SWAYING OF THE BODY

People use their eyes, inner ears, and proprioceptors in muscles, tendons, and joints to maintain balance while standing. As the body sways ever so slightly from front to back or side to side, receptors from these senses direct signals to the brain, informing it of the change. The brain responds by commanding muscles to contract to maintain the correct posture. In other words, *feedback* from our sensory receptors repeatedly corrects small deviations in posture.

If a person has damage to one sense, the other senses continue to support balance. For example, a person with damage to the sense of proprioception uses vision to stay balanced. But if you ask this person to shut both eyes, the body will sway excessively.

To test balance, ask a subject to stand at attention—with the feet together and the body still. Is there any swaying from side to side or front to back? Now ask the subject to continue standing but to shut both eyes. Does the swaying increase? Then ask the subject to stand on one foot while keeping both eyes shut. Does the swaying increase further?

Repeat these procedures on yourself to experience the feelings of your subject.

WALKING A STRAIGHT LINE

Ask your subject to walk along a straight line, one that you have drawn or that is already present. Does the subject sway or stagger? If so, the eyes, proprioceptors, or parts of the brain to which they connect might be defective. Now ask your subject to close his or her eyes and again walk the straight line. What happens?

Materials

- Frog
- Large beaker or similar container
- A turntable, such as a rotating stool or rotating spice tray (optional)

ROTATION OF A FROG

Place a frog in the center of a large beaker or similar container. If you have a turntable, place the beaker at its center. Rotate the container 90 degrees to the left, then 90 degrees to the right. Watch the frog move its head and sometimes its entire body in a direction opposite to the rotations (FIG. 23-2). It does this to regain its normal position.

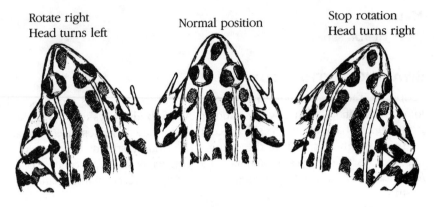

Rotate right
Head turns left

Normal position

Stop rotation
Head turns right

23-2   Responses of a frog to rotation

As you turn the frog, fluid moves backward through its semicircular canals, bending the hairs of its sensory receptors. The receptors direct signals to the brain, informing it of the turn. In response, the brain commands muscles of the body to compensate.

Now rotate the frog around and around continuously at a constant rate. The fluid in its canals flows backward at first, but soon catches up, moving at the same speed as the animal. The hairs of the balance receptors are no longer bent, so the frog returns its head to its normal position.

Stop the rotation suddenly. The fluid—which was moving as fast as the frog—continues to flow. Which way does the frog now turn? If the rotation was to the right, the fluid continues flowing to the right when the animal stops moving. The frog's brain responds to the rightward bend of the hairs in its canals by directing the frog also to turn right.

Materials

- Pepper
- Large beaker or other circular container of water
- A turntable (optional)

ROTATION OF WATER

Half fill a beaker or other circular container with water. Sprinkle pepper on the water. If you have a turntable, such as a rotating stool or spice tray, set the beaker at the center of it. Watch the water and pepper as you start to turn the beaker clockwise. The particles stay where they are. In relation to the beaker, however, they move counterclockwise.

Continue rotating the beaker clockwise at a constant speed. Eventually the water and the pepper move clockwise at the same speed as the beaker.

Abruptly stop the rotation. The water and pepper continue to move clockwise until molecular friction stops them.

The movements of fluid in the beaker resemble the movements of fluid in the semicircular canals of frogs, humans, and other animals.

Materials

- Human subject
- Rotating chair or stool
- A large paper bag

ROTATION OF A HUMAN

The semicircular canals of humans and frogs react to gravity in much the same way. Prove this by standing and turning quickly round and round eight or ten times. Then stop. Which way do you lean? Like the frog, you recognize body position by the flow of fluid and the activation of receptors in your semicircular canals. And like the frog, if you started by turning to the right, you will now be falling to the right.

After recovery, repeat the rotation. Shut your eyes after eight or ten revolutions, and quickly raise your index finger perpendicular to the ground. Have a friend watch. The task is not as easy as it seems. The finger goes to one side just as does the body.

Have your friend turn around quickly eight or ten times, then stare at your nose. Watch his or her eyes. They drift sideways, then jerk back, repeatedly. This movement is called *nystagmus*. The jerking of the eyes occurs also during rotation, allowing the subject briefly to fix the eyes on surrounding objects. Ballet dancers similarly and purposely jerk their heads into fixed positions when they rotate.

For a last test, you need a chair or stool that you can rotate. Have your friend sit on the chair or stool, hands gripping the armrests or rim, feet slightly off the floor. Cover his or her head with a paper bag. Rotate the chair or stool eight or ten times, and let it coast to a stop. Have the subject say when it stops. Because the fluid in the semicircular canals continues to rotate after the stop, the subject might think the chair or stool is still rotating.

OTHER ACTIVITIES

Crayfish have long "feelers," called *antennae*, and short ones, called *antennules*, projecting from their heads. Each antennule has a cavity at its base called a *statocyst*, an organ of equilibrium similar to the utricle and saccule in humans. Grains of sand lie in the statocysts. Gravity pulls on these grains, bending hairs of receptor cells, providing sensations of position.

When crayfish molt, they shed their external skeletons, including the statocysts. Then the skeleton and statocysts grow back. Each statocyst has an opening to the exterior through which the crayfish scoops in new sand, allowing its organ of equilibrium to work again.

Researchers have tricked crayfish by sprinkling iron filings in their aquaria to replace the sand. After molting, the crayfish pick up the filings in their statocysts. The animals continue to crawl about normally until someone brings a magnet nearby. The magnet attracts the filings in the statocysts, causing the hairs of receptor cells to be bent, causing the crayfish to swim upside down or sideways. If you know a stream in which to catch crayfish, try this experiment.

Part 7

Bones and Muscles

Chapter **24**

Anatomy
of a skeleton

What does the skeleton do for us? More than most people imagine. First and most evident, it *supports* the soft tissues of our bodies. This support allows us to sit, stand, walk, and run—to be physically active.

Second, the skeleton *protects* our vital organs by enveloping them. The ribs protect the heart and lungs; the pelvic bones protect reproductive organs; and the skull protects the brain. In the skull, for example, the respiratory centers are buried in the brain's medulla. Depth of protection is important because damage to these centers prevents breathing.

Third, the skeleton *provides leverage* to move the body and external objects. When walking, for example, we contract the gastrocnemius muscles in our calves. Each gastrocnemius attaches by a sturdy tendon to a calcaneus (heel bone). When we contract a gastrocnemius muscle, it pulls the calcaneus upward, forcing the ball of the foot downward, allowing us to step.

Fourth, the skeleton *makes red and white blood cells and platelets.* Specifically, the red marrow in bones of the skull, chest, backbone, and pelvis makes blood cells to replace others that break down. Then the cells and cell fragments, the platelets, enter tiny vessels that lead out of the bones. The red cells carry oxygen to all tissues, sustaining life; the white cells destroy invading microbes, eliminating infections; and the platelets cause blood to clot, stopping blood loss.

Fifth, the skeleton *stores and releases calcium*. About 99 percent of all calcium in the body is in bones, and there is a continuous exchange of this calcium with blood. In other words, the bones are a reservoir for calcium that is released to the blood for transport to other parts of the body. Blood adds calcium to bones in times of plenty and withdraws it in times of need.

The body needs a steady supply of calcium to regulate muscular contractions. If the calcium level becomes too low, the muscles contract excessively, causing stiffness and tremors. Such contractions might even interfere with breathing. If instead the calcium level becomes too high, the muscles relax, causing weakness.

Materials

- Figures 24-1 and 24-2
- A mirror
- A human skeleton (optional)

EXAMINING THE SKULL

Because we each carry our own skeleton, we have an excellent model for study. Feel your bones and compare them with the human skeleton of FIGS. 24-1 and 24-2. Also compare them with a real human skeleton, if available.

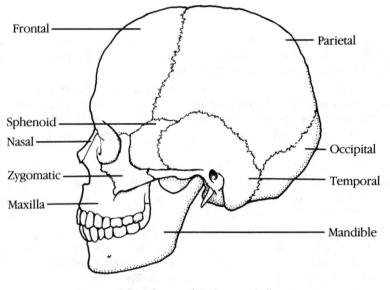

24-1   Bones of the human skull

Begin with the skull. It is formed by many bones, tightly bound at interlocking *sutures*. The front bone—called the *frontal* bone—is followed by *parietal*, *occipital*, and *temporal* bones. Feel each bone.

The occipital bone, at the back and base of the skull, has a large opening, a *foramen magnum*, through which the brain connects to the spinal cord. It also has a rough area at the back where muscles of the neck attach. These muscles help balance the head on the backbone. Feel the muscles and their attachment to the skull.

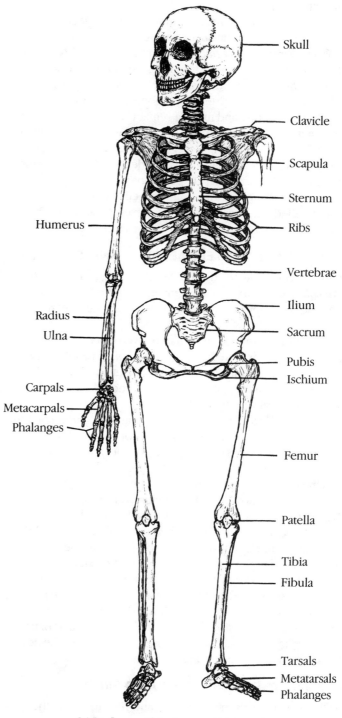

Skull

Clavicle

Scapula

Sternum

Ribs

Humerus

Vertebrae

Ilium

Radius

Sacrum

Ulna

Pubis

Ischium

Carpals

Metacarpals

Phalanges

Femur

Patella

Tibia

Fibula

Tarsals

Metatarsals

Phalanges

24-2 Bones of the human skeleton

The temporal bone occupies the temple at the side of the head. Notice the tubular opening into the bone, an *auditory canal* that channels sound to the hearing apparatus. Stick your finger into your ear to feel the canal.

Look also for the *zygomatic* bone of the cheek, the *sphenoid* bone above and behind it, the *nasal* bone of the nose, the *maxilla* bone of the upper jaw, and the *mandible* bone of the lower jaw. Move your jaws as you feel the joint at which the mandible joins the temporal bone. Notice that the mandible moves while the temporal bone and maxilla remain still.

Look in a mirror at your teeth. Choosing either the upper or lower jaw, notice the four, middle-most, chisel-shaped *incisor* teeth, used for cutting. Follow backward from an incisor of one side to a fanglike *canine* tooth, used for tearing. Dogs have unusually large canines. Continue backward to the two *premolar* teeth and two or three *molar* teeth, used for grinding. The third molar, also called the "wisdom tooth," develops at 17 to 25 years of age. Surgeons sometimes remove wisdom teeth that crowd other teeth.

EXAMINING THE VERTEBRAE, RIBS, AND STERNUM

There are 33 *vertebrae* in the human skeleton, forming a long, S-shaped *vertebral column* (backbone). The curvature of this column sometimes varies through inheritance or disease. In scoliosis, for example, the column bends sideways, and in kyphosis, it becomes hunched.

The vertebral column is divided into seven *cervical* vertebrae in the neck, twelve *thoracic* vertebrae in the chest, five *lumbar* vertebrae at the waist, five sacral vertebrae fused into a single *sacrum* at the hip, and five coccygeal vertebrae fused into a *coccyx* at the tail. Cats and other animals with long tails have many coccygeal vertebrae.

Each vertebra has a large central opening. The *spinal cord* runs through here in a *vertebral canal* formed by a series of vertebral openings.

Each vertebra also has a *spine* projecting outward at the back. Bend your body at the waist and feel along the vertebral column. The spines feel like bumps along the midline, one bump for each vertebra. Muscles attach to these spines, muscles that allow you to straighten your back.

Each vertebra is separated from the next by a cartilaginous *disk* and bound to the next by *ligaments*. Individually the ligaments bind tightly, but collectively they allow movement of the vertebral column. Bend your body to the front and back and to the left and right to find the extent of this movement.

Now feel the *sternum* (breastbone) at the midline of your chest, and the *ribs* curving around the sides of your chest. There are twelve ribs enclosing the lungs and heart. Inhale and exhale deeply to feel the movements of these ribs.

The heart, about the size of a fist, is located under and partly to the left of the sternum. Place your hand at this position. Can you feel the heartbeat? If not, run in place for 1 or 2 minutes and try again.

EXAMINING THE SHOULDERS AND HIPS

The girdles are bones of the shoulders and hips to which the arms and legs attach. The arms attach to a *pectoral girdle* and the legs to a *pelvic girdle*.

The pectoral girdle consists of two *clavicles* or collarbones and two *scapulas* or shoulder blades. Feel a clavicle, a slightly curved bone passing from the upper end of the sternum to the shoulder. People sometimes fracture it while using their outstretched hands to break a fall. Feel also the scapula of your back or of a friend's back. Shrug your shoulders to find the range of movement of your clavicles and scapulas.

Move downward to your hip to feel the tightly bound *pelvic* bones. The crest of each bone curves from side to back at your waist, leading to the sacrum. Now sit, if you are not already seated. You are sitting on your pelvic bones.

EXAMINING THE ARM

Feel the *humerus*, the bone of the upper arm. Its upper end has a round head that fits into a socket. Swing your arm to feel the movement.

The humerus connects below to two long bones: the *ulna* and *radius*. With your free hand, feel the ulna from your elbow to your wrist and the radius as far as you can. The radius is on the thumb side of the forearm. Rotate your forearm to feel movements of the radius and ulna.

Touch your fingers to the humerus and ulna of the other arm. Now flex (bend) and extend the arm to feel how the bones work together as a hinge.

Ligaments bind eight small bones—the *carpals*—in the wrist. Bend your wrist from front to back and side to side to feel the range of movement.

Feel the five *metacarpal* bones in the palm of your hand. They connect the carpals to the *phalanges*, the bones of the fingers. Wiggle the phalanges.

Your thumb is said to be *opposable* because you can hold it opposite the other fingers. Because it is opposable, you can grip and use tools. Most other animals lack opposable thumbs, making them less dexterous.

EXAMINING THE LEG

Feel the *femur*, the bone of the thigh, covered by bulky muscles. Its upper end has a round knob that fits into a socket in the pelvic bone. Rotate the leg at this socket.

The femur connects below to a large bone, the *tibia* (shinbone). Feel its ridge along the front of the leg. Beside the tibia, along the outer margin of the leg, is another, more narrow bone, the *fibula*. Feel it also, starting near the knee.

The knee itself is formed by the *patella*, a knobby bone bound to the tibia by a patellar ligament. Place your hand over the patella, touching the femur and tibia with your palm and fingers. Now flex and extend your leg to feel how the bones work together as a hinge.

Ligaments bind seven bones—the *tarsals*—in the ankle. Bend your ankle front to back and side to side to feel the range of movement.

Feel the five *metatarsal* bones at the top of your foot. They connect the tarsals to the *phalanges* or toes. In contrast to the fingers, the toes of humans are unopposable and of little use in handling tools.

REVIEW OF BONES

Do you remember the names of the many bones? To find out, wiggle your phalanges, put your metacarpals on your patella, and rotate your humerus. Then think of other ways to use bones.

Materials

- Chicken bones
- Vinegar
- Cup or beaker
- Oven

COMPONENTS OF BONE

The bones have two major components: *collagen* and *calcium phosphate*. The collagen of bone is a protein, the same protein found in the cartilage of the ears. The calcium phosphate is part of a mineral called *hydroxy-apatite*.

To feel the properties of collagen, bend your ear. If you imagine this flexible collagen being reinforced by the deposition of hard hydroxyapatite, you will have some idea of the nature of bone. About half the volume of bone is collagen and the other half, hydroxyapatite.

To separate the components of bone, place one or two chicken bones—preferably to include a wishbone—in a cup of vinegar. Vinegar is 5 percent acetic acid. The mild acid dissolves the mineral of the bones in a few days, leaving behind the collagen. What effect does this have on the bones? Try bending them to find out.

Also place one or two bones in an oven at 121° C (250° F) for 3 hours. Heat dries the collagen. Try to bend these dry bones of hydroxyapatite, then break them. Notice how brittle they are.

Vitamin D promotes the absorption of calcium from the digestive tract and its deposition in bones. In rickets, vitamin D is deficient. Why do the victims of rickets have bowed legs?

Materials

- Dead animal
- Surgical gloves
- Knife
- Jar or bucket
- Brush
- Borax
- Hydrogen peroxide
- Wire

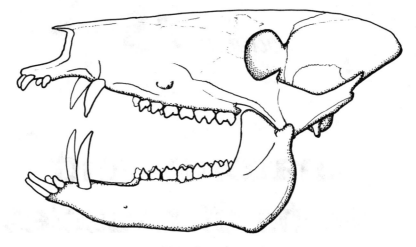

24-3   Javelina skull

PRESERVATION OF SKULLS AND SKELETONS

If you appreciate the beauty and instructional value of skulls and skeletons, prepare one or more of your own (FIG. 24-3). Start with the head of some animal you have bought for meat. For the latter, check the yellow pages of your phone book under "Meat." A custom butcher or meat packer might provide skulls not readily available at your grocery.

CAUTION: Avoid partially decomposed animals found in nature. Some carry disease. If you want such bones, collect the bones after the meat entirely decomposes, freeing the bones to bleach in the sun.

Wearing surgical gloves, remove the eyes, brain, and as much flesh as possible. You might want to save the eyes and brain for further study (chapters 16 and 20). Put the cleaned skull in water outdoors where the remaining meat will decompose in one to four weeks. If you want to hurry the process, add meat tenderizer, or, better yet, boil the skull. Then rinse it and brush off loose particles. Next put the skull in a cleaning solution prepared by dissolving 50 grams (⅕ cup) of borax in 1 liter (1 quart) of water. Let it soak overnight, scrubbing occasionally, to remove traces of fat. Rinse and then bleach the skull in 3 percent hydrogen peroxide for 10 hours to get an attractive, durable, gleaming white skull.

Preparation and mounting of whole skeletons is trickier. Use the same cleaning procedure but watch the ligaments. You must stop the decomposition at just the right time to preserve them. Then use wire to mount the skeleton in a natural position.

OTHER ACTIVITIES

Look at X-rays of skeletons. This will allow you to relate the bones to overlying tissues and to see fractures, if there are any. Can you identify the bones?

Chapter **25**

Anatomy and actions of muscles

*H*umans have over 600 skeletal muscles, comprising 40 to 45 percent of our body weight. Some are large muscles that we all recognize as muscles, such as the biceps that flexes the arm. Others are small muscles, such as the rectus muscles that turn the eyeballs.

Each muscle contains hundreds to thousands of narrow, threadlike cells called *muscle fibers*. The cell membranes of the fibers join sheaths of connective tissue, and the sheaths join *tendons* that attach to bones. When the muscle fibers contract, they pull the sheaths that move the tendons that move the bones.

A skeletal muscle is usually anchored at one end to an immovable or slightly movable bone or bones, a junction called the *origin*. The same muscle is usually bound at the other end to a movable bone or bones, a junction called the *insertion*. The biceps of the upper arm, for example, originates on the fixed scapula bone of the shoulder and inserts on the movable radius bone of the forearm. The joint between the upper arm and forearm acts as a hinge. This arrangement allows the biceps, inserted near the hinge, to pull the radius and the hand quickly through a long distance.

There are two other kinds of muscle: cardiac and smooth. *Cardiac muscle* is in the heart. Its fibers are long and thin, like those of skeletal muscle, but different because they branch and fuse with each other. *Smooth muscle* is in the digestive tract, respiratory tract, blood vessels, and other internal organs. Its tapering cells are smaller than those in skeletal and cardiac muscle, and contract more slowly.

Materials

- The muscles of your body
- A mirror (optional)

WINKING WITH YOUR ORBICULARIS OCULI

The orbicularis oculi is a circular muscle that runs around each of the eyes (FIG. 25-1). We contract it to wink or close our eyes.

To feel the orbicularis oculi, place the fingers of your hand firmly around the outer margins of the paired eyelids, above the cheekbone and under the eyebrow. Now wink the eye tightly and keep it closed. Do you feel the tension of the contracting muscle fibers? Being circular, the muscle squeezes toward the center as it contracts.

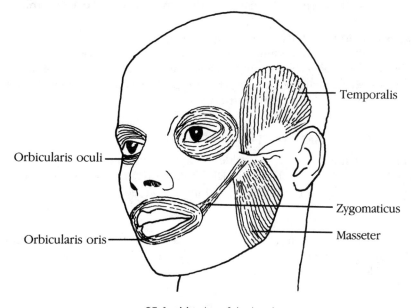

25-1   Muscles of the head

You might also want to look in a mirror as you contract the orbicularis oculi and other facial muscles that follow.

KISSING WITH YOUR ORBICULARIS ORIS

The orbicularis oris is a circular muscle that runs around the mouth, as the orbicularis oculi runs around the eyes (FIG. 25-1). We contract the oris to kiss or whistle.

Put the fingers of one hand around your mouth near the lips. Then pucker up for a pretend kiss or a real whistle. Do you feel the change in tension as you pucker? Like the orbicularis oculi, the orbicularis oris squeezes centrally as it contracts.

SMILING WITH YOUR ZYGOMATICUS

The zygomaticus is a straplike muscle that originates on the zygomatic bone (cheekbone) and inserts at the corner of the mouth (FIG. 25-1). We contract it to smile.

Put the fingers of your hand on the zygomaticus muscle in your cheek. Now give a big, big smile. Do you feel the muscle contract? We should exercise this muscle often.

CHEWING WITH YOUR MASSETER AND TEMPORALIS

The masseter muscle originates on the zygomatic bone (cheekbone) and inserts at the back of the mandible, or jawbone (FIG. 25-1). We contract it to chew food and grit our teeth.

Press your fingers on the masseter while chewing gum. Feel it contract as you raise your jaw.

The temporalis muscle also helps with chewing. It originates on the temple, at the side of the skull, and inserts on the mandible (FIG. 25-1).

Press your fingers on the temporalis as you continue chewing the gum. Do you feel both it and the masseter contract as you raise your jaw? Grit your teeth to feel the muscles contract more strongly.

RAISING THE ARM WITH YOUR DELTOID

The deltoid muscle of the shoulder is a frequent site for intramuscular injections. It originates on the clavicle (collarbone) and scapula (shoulder blade), passes across the shoulder, and inserts on the humerus bone of the upper arm (FIG. 25-2). We contract the deltoid to abduct the arm—that is, to raise it outward and upward.

Press one hand against the shoulder and upper arm of the opposite side of the body. Raise this arm to the side until it is higher than your shoulder. Do you feel the deltoid stiffen as it contracts?

FLEXING THE ARM WITH YOUR BICEPS

The biceps brachii muscle, at the front of the upper arm, originates on the clavicle (collarbone) and scapula (shoulder blade) and inserts on the radius bone of the forearm (FIG. 25-2). We contract the biceps to bend the arm at the elbow. Some muscular men also contract it to show off.

Start this trial with your arm fully extended and your opposite hand on the biceps. Now flex the arm. Do you feel the biceps contract? Grip the muscle again as you lift a heavy weight. Do you feel a stronger contraction—that is, an activation of more muscle fibers?

EXTENDING THE ARM WITH YOUR TRICEPS

The triceps brachii muscle, at the back of the upper arm, originates mainly on the humerus bone of the upper arm and inserts on the ulna bone of the elbow (FIG. 25-2). We contract the triceps when we extend (straighten) the arm.

Start this trial with your arm flexed (bent) and your opposite hand on the triceps. Now fully extend the arm and keep it extended. Do you feel the triceps contract?

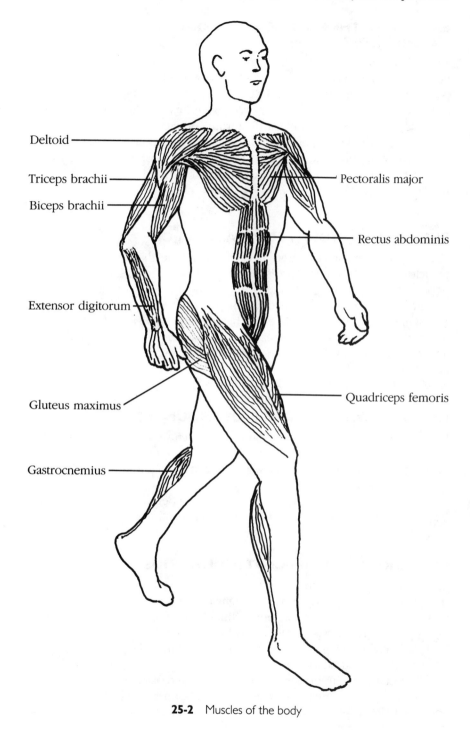

Deltoid

Triceps brachii

Biceps brachii

Extensor digitorum

Gluteus maximus

Gastrocnemius

Pectoralis major

Rectus abdominis

Quadriceps femoris

25-2   Muscles of the body

EXTENDING THE FINGERS WITH YOUR EXTENSOR DIGITORUM

Turn to the back of your forearm. The extensor digitorum originates on the humerus and inserts by four long tendons on the four fingers (FIG. 25-2). We contract it to extend the fingers.

Feel the back of the forearm as you extend its four fingers upward and outward. Do you feel the extensor digitorum contract?

As you extend your fingers, notice that the four tendons of the extensor digitorum show prominently above the knuckles. Feel one of these long, narrow tendons, following it from the muscle, where it originates, to the finger, where it inserts.

PULLING WITH YOUR PECTORALIS MAJOR

The pectoralis major is a large, fan-shaped muscle that originates on the clavicle, ribs, and sternum (breastbone) and inserts on the humerus of the upper arm (FIG. 25-2). We contract it to draw the arm toward the chest, as in swimming, throwing, pulling, and climbing.

While sitting in a chair, place one hand on the opposite side of the chest where the pectoralis major crosses the armpit. Move your arm outward, then inward, pressing on the armrest. With which movement do you feel the pectoralis contract?

EXHALING WITH YOUR RECTUS ABDOMINIS

Proceed to the midline of the abdomen. The rectus abdominis is a straplike muscle that originates on the pubic bone, near the crotch, and inserts on the sternum and lower ribs (FIG. 25-2). We contract it to exhale air forcefully, strain at the stool, and do sit-ups.

Place your hand on the rectus abdominis, just below the sternum. Exhale *all* of your air forcefully. Do you feel the muscle contract?

CLIMBING WITH YOUR GLUTEUS MAXIMUS

The gluteus maximus is the largest muscle in the buttocks, a site sometimes used for intramuscular injections. It originates on the pelvic bone (hipbone) and sacrum and inserts on the femur (thighbone) (FIG. 25-2). There are two gluteus maximus muscles, one for each hip. We contract them to thrust our legs backward, as in running, springing, and climbing. Other, deeply placed gluteus muscles are more used for walking.

To feel the gluteus maximus muscles, stand and place both hands behind you on the buttocks, near the two thighs. Walk up a flight of stairs as you continue to feel the buttocks. Notice that a gluteus maximus contracts for one leg, then the other, as you step up.

KICKING WITH YOUR QUADRICEPS FEMORIS

The quadriceps femoris is a group of four muscles in the front of the thigh. These muscles originate on the femur and pelvic bone and insert by a common tendon on the patella, or kneecap (FIG. 25-2). The tendon continues over the patella as a ligament that connects the patella to the tibia (shinbone). (By definition, tendons bind muscles to bones; ligaments bind bones to bones.) We contract the quadriceps femoris to extend the leg.

Place your hand on the thigh near the knee. Extend your leg forcefully, as in kicking, and keep it extended. Do you feel the quadriceps contract? Run your fingers down the leg to identify the quadriceps tendon, patella, patellar ligament, and tibia. Contract the quadriceps again to feel the tendon and ligament tighten.

STANDING WITH YOUR GASTROCNEMIUS

Shift to your calf muscles. The gastrocnemius originates at the lower end of the femur and inserts by a tendon of Achilles on the calcaneus, or heel bone (FIG. 25-2). We contract the gastrocnemius to stand, walk, and run.

Sit on the floor. Place your hands on the gastrocnemius muscles of your two calves to feel how slack they are.

Now stand. While bending at your waist, reach down to feel the gastrocnemius muscles again. Have they contracted to keep you standing? As you continue to feel the muscles, stand on your toes. Have the muscles contracted more forcefully yet?

REVIEW OF MUSCLES

Do you remember the names of the many muscles? To find out, name the contracting muscles as you wink, kiss or whistle, smile, chew, raise your arm, flex your arm, extend your arm, extend your fingers, draw your arm toward your chest, exhale, step up, kick, and stand.

Materials

- Chicken dinner (preferably to include a whole chicken)

OBSERVING THE MUSCLES YOU EAT

When you next eat chicken or other meat, notice the structure of the muscles. Can you find or imagine where the muscles originate and insert? Are tendons present? Are there blood vessels to provide nutrients and oxygen to the muscles? Can you tell what movements the muscles control?

In chickens, the dark meat contains many mitochondria—the dark, subcellular structures that generate ATP and energy. Dark meat is in the legs, because chickens walk a lot; their legs need many mitochondria to supply energy. This is especially evident in barnyard chickens.

In contrast, the white meat has few mitochondria. White meat, therefore, is in the flight muscles of the breast. Because chickens seldom fly, the flight muscles need few mitochondria to supply energy. In contrast, what color would you expect breast meat to be in migrating birds?

OTHER ACTIVITIES

Check out a human anatomy book. It will have drawings or photographs of the muscles you studied and many more. Identify additional muscles and devise ways to examine their contractions.

You will feel stronger contractions if you contract the muscles against resistance. For example, in testing the trapezius muscle at the back of the neck, you should press your head backward against your hand or a headrest as you feel the muscle with your other hand.

Chapter **26**

Contraction
of muscles

As a car needs gasoline, our bodies need glucose. Carried to the muscles by blood, the glucose and other nutrients combine with the oxygen we breathe, forming *ATP* (adenosine triphosphate). ATP is the fuel that supports work—that powers the contraction of muscles.

When nerve impulses enter muscle cells (muscle fibers), the impulses cause the release of energy from ATP. This energy allows the filaments inside the muscle fibers to slide toward each other, producing the contractions.

During the delay between an impulse and the start of contraction, ATP releases its energy, filaments slide, and connective tissue around the muscle stretches. This stretch dissipates energy. Thus a single, initial stimulus to the muscle yields a weak twitch. But if other stimuli quickly follow the first, ATP splits repeatedly, allowing little time for the muscle to relax. Contractions, therefore, connect and summate; their united strength leaps beyond that of a single twitch.

Hundreds of nerve cells carry impulses to tens of thousands of muscle fibers in each muscle. Sometimes the brain discharges only a few of the nerve cells, producing weak contractions. Other times the brain discharges many nerve cells, producing strong contractions. The contractions are stronger when many nerve cells activate each muscle and when each nerve cell discharges frequently.

When many nerve cells and muscle fibers discharge frequently, ATP is consumed rapidly. Glucose then breaks down, releasing its energy to form more ATP. If oxygen is present, glucose breaks down completely to carbon dioxide and water, providing much energy to form ATP; each molecule of glucose produces 36 molecules of ATP. But if oxygen is absent, glucose is converted into lactic acid, providing little energy to form ATP; each molecule of glucose produces only 2 molecules of ATP.

Because ATP is the only direct source of energy that muscle can use, the availability of glucose and oxygen affects the extent and duration of contractions. Also, the circulation of blood to muscles affects exercise, since the blood carries glucose and oxygen.

When we walk or do other light exercise, our bodies have ample time to deliver oxygen to contracting muscles; we form nearly all the ATP possible. The abundant ATP allows us to continue walking for many hours. But when we sprint or do other strenuous exercise, our bodies fail to deliver enough oxygen to the muscles; we form inadequate ATP. The scarce ATP causes us to slow or stop sprinting in one minute or less. We cannot provide enough oxygen and ATP to allow continued contraction of muscles.

Materials

- A stethoscope and alcohol, or another human subject

SOUND OF CONTRACTING FIBERS

Locate the masseter muscle at the back of the jaw by gritting your teeth (FIG. 25-1). With one hand, feel it harden and bulge. This is a muscle that elevates the jaw in chewing.

Now clean the earplugs of a stethoscope with alcohol, and insert them in your ears. Place the stethoscope disk against the masseter. If you have no stethoscope, place your ear directly against the masseter of another person. If the room is quiet, you will hear the asynchronous murmur of groups of muscle fibers that maintain muscle tone. These are the few fibers that keep the jaw elevated at rest. Different groups of them alternately contract and relax.

While still listening with the stethoscope, firmly grit your teeth. How does gritting change the sound? How does the new sound relate to the number of stimuli entering muscle fibers, the rate of fiber activation, and the number of fibers contracting?

Materials

- Spring-operated clothespin
- Pencil and notebook
- A human helper (optional)

FATIGUE OF CONTRACTING MUSCLES

When we contract muscles repeatedly at a fast rate, the muscles become fatigued. In sprinting or fast running, for example, we become exhausted in seconds or minutes, then slow or stop. We fatigue because we cannot make ATP quickly enough to support quick contractions.

To demonstrate fatigue, grip the two wooden prongs of a spring-operated clothespin. Hold them between your thumb and index finger. Now squeeze the prongs together, causing the other end to open (FIG. 26-1). Fully open and

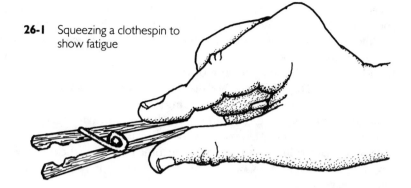

26-1   Squeezing a clothespin to show fatigue

close the clothespin as rapidly as possible for 90 seconds. Have a friend count and record the number of contractions you make from 0 to 30, 30 to 60, and 60 to 90 seconds. If you have no helper, stop just long enough at the end of each 30 seconds to record the number of contractions, then continue. Do you slow as the effort continues? Do you have increasing difficulty in forcing the prongs fully open?

The contracting muscles consume more glucose, oxygen, and ATP than usual. As they do so, the blood flow to them increases, bringing more glucose and oxygen, but not enough to compensate for the increased consumption. Thus the production of ATP falls behind, and the muscles weaken. They cannot contract as rapidly or as forcefully as they did at the beginning of exercise.

Fatigue also results from an increase of lactic acid at the site of contraction. Lactic acid and other acids promote the binding of calcium ions to the sarcoplasmic reticulum, a network of tubules within the muscle fibers. This binding hinders the action of ATP and thereby the sliding of muscle filaments.

In squeezing the clothespin, you likely used the fingers of the hand with which you customarily write and do other activities. The muscles in this hand and arm are better developed than in the opposite hand and arm because you exercise them more. Also, they are better coordinated.

Now grip the clothespin with the index finger and thumb of the hand you seldom use. Squeeze the prongs together as rapidly as possible for 30 seconds. Record the number of contractions. How does this number compare with that from 0 to 30 seconds for the original hand?

Materials

- Pencil
- Sphygmomanometer cuff or tourniquet
- A human helper

BLOOD SUPPLY AND FATIGUE

To relate the circulation of blood to the contraction of muscles, you can shut off the circulation as you work—for example, as you operate the clothespin used earlier, or as you handwrite this paragraph. Begin by wrapping the cuff of a sphygmomanometer (the apparatus used to measure blood pressure) or a tourniquet around your upper arm and fastening it in place. Then copy this paragraph in a notebook. After one minute, while you are still writing, have a friend inflate the cuff to a pressure of 150 mm Hg and hold it there or tighten the tourniquet. The inflated cuff or tightened tourniquet stops the blood flow to your hand. Mark the place in your writing where the circulation was stopped. Continue writing for one or two minutes at your normal speed. Does your handwriting deteriorate? Do you slow down or use muscles that you seldom use for normal writing? Does your hand become pale? If so, the red blood cells have released their oxygen, causing them to lose color. How do blood flow and oxygen relate to fatigue?

When your hand is too tired to continue writing, stop. Have your friend *immediately deflate the cuff or loosen the tourniquet.* What happens now to the color of the hand that was writing? Is it redder than the other hand? What does this suggest about the difference in blood flow?

Normally there is enough oxygen in blood and the overlying skin to make them red. In the skin, this color shows best in people who have little melanin (skin pigment). The redness results from a combination of oxygen with hemoglobin in the red blood cells. But when the cuff is inflated as a person continues writing, the contracting muscles consume virtually all the oxygen from the blood. It loses its red color.

In addition, the lack of oxygen and associated accumulation of lactic acid in the blood causes the blood vessels of the hand to *dilate*—that is, causes their diameters to increase. Dilation occurs in part because the lack of oxygen curtails the formation of ATP. Without ATP to cause contraction, the muscle in the walls of blood vessels relaxes. Consequently, the vessels dilate and stay dilated even when blood is readmitted to the hand. While the vessels are dilated, blood rushes in to fill them and makes the hand appear red. The dilation and redness continue until oxygen generates enough ATP to permit the contraction of the vessel walls.

Allow the color of your hand to return to normal. This indicates that the blood and oxygen in the muscles are restored. Then try handwriting again. Does the writing appear normal?

OTHER ACTIVITIES

Young mice, rats, and I suspect, other small mammals can be trained to balance and walk on a rotating rod. They continue to do this until they become tired, at which time they drop to the floor beneath them. You can use such a system to test the effects of fatigue, sleep, cola drinks, and so on.

To construct the apparatus, obtain a wooden dowel that is 2.5 centimeters (1 inch) or wider in diameter. Cut two cardboard disks about 20 centimeters (8 inches) in diameter with holes at their centers to fit the dowel. Slip the disks on the dowel, allowing enough space between them for your animal to walk. Mount the dowel on a support that will allow it to rotate. Then connect a pulley from a small electric motor to the dowel. Adjust the motor and the apparatus to turn the dowel slowly. The rate of turning should allow the animal to walk or slowly run and stay balanced. Train the animal for several days or weeks to continue walking or running on the rod as it rotates. Then try your experiments.

Part 8

Oxygen, endocrines, and digestion

Chapter 27

Oxygen and heat

Our bodies need energy to move muscles, transmit nerve impulses, form biochemicals—to keep things working. This is why we eat. The nutrients of food break down, releasing carbon dioxide, water, and energy.

We use most of this energy to make ATP. The ATP later breaks down, releasing its energy to support movement and other activities. We release other energy as heat—a byproduct of the breakdown of nutrients and ATP. It keeps us warm and provides optimal temperatures for our metabolic reactions.

Metabolism is the chemical process by which substances are assembled or broken down. In this process, energy can be either consumed or released. Here we shall concentrate on the breakdown of substances during which energy and heat are released and oxygen consumed.

Biologists sometimes measure the heat production of humans or other animals by placing them in closed metabolic chambers. Water circulates through pipes surrounding each chamber, where heat from the animal warms it. Sweat is absorbed and weighed. By knowing the temperature increase of the water and the extent to which sweat cools the body, biologists can calculate the metabolic heat produced.

Alternatively, biologists calculate the heat produced by measuring the oxygen consumed. The animal breathes into and out of a closed container of oxygen, consuming oxygen. As it inhales oxygen, it exhales carbon dioxide into an absorbent, such as soda lime. The only net change, therefore, is the removal of oxygen. The consumption of each liter of oxygen generates 4.8 kilocalories of heat. (When referring to food, most people use the word "calorie" rather than kilocalorie. In this sense, the two words have the same meaning.)

Exercise raises the metabolic rate. An adult human produces about 1600 kilocalories of heat per day when resting. This is called the *basal metabolism* or basal heat production. It is measured for a short time while the person lies in a quiet room after an overnight fast, then multiplied by a factor that gives total heat per day. Sitting raises the heat production 25 percent, walking 300 percent, and running 500 percent or more. The amounts of heat we produce and food we consume depend mainly on our levels of activity.

Materials

- Mouse or small rat
- 1-liter or 1-quart widemouthed bottle
- 1-hole rubber stopper to fit the bottle
- Soda lime
- Surgical gloves
- Graduated 1-milliliter plastic pipette
- Screen wire or small-mesh hardware cloth
- Adhesive tape
- Soap solution (such as that used by children to blow bubbles)

MEASURING THE OXYGEN CONSUMPTION OF A MOUSE OR RAT

Select a glass or plastic widemouthed bottle having a 1-liter (1-quart) or greater capacity (FIG. 27-1). If your experimental animal is a rat that weighs 100 grams or more, use a two-liter bottle. Pour granules of soda lime, a carbon dioxide absorbent, into the bottle to cover the bottom to a depth of about 1 centimeter (⅓ inch).

CAUTION: Soda lime is alkaline. Wear surgical gloves while handling it, and wash your hands afterward.

Turn the bottle on its side, and unless it is square-sided, brace it to keep it from rolling. Get a 1-hole rubber stopper that fits its mouth. Insert the end of a graduated 1-milliliter pipette into the stopper. *For safety, choose a plastic pipette, not glass.* Wet the tip of the pipette to make it easier to insert.

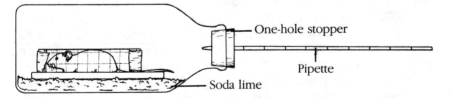

27-1   Apparatus for measuring oxygen consumption

Construct a wire cylinder large enough to hold the mouse or rat but small enough to prevent it from turning around. Screen wire or small-mesh hardware cloth makes a sturdy enough cylinder. Close one end of the cylinder but leave the other end open. Tape the completed cylinder to a piece of cardboard that you will later place on top of the soda lime in the bottle.

CAUTION: Have an experienced person, such as the owner of a pet store, show you how to handle the mouse or rat gently and safely.

When ready, direct the mouse or rat toward the opening of the cylinder. Do not force it to enter. Most rodents voluntarily dart into holes of this kind. When the animal is inside, immediately tape the opening shut. Now rinse the inside of the pipette with water, place the rodent and its cylinder in the bottle, and plug the rubber stopper firmly into the mouth of the bottle.

While the animal is still inspecting its cylinder, dip your finger into a soap solution—the same as that used by children to blow bubbles—and wipe it across the end of the pipette. The resulting soap film will travel slowly down the tube as the animal consumes oxygen. The soda lime prevents the collection of carbon dioxide that would stop the bubble's descent. Record the time for the bubble to move from the 0-milliliter to the 1-milliliter calibration line. If the animal is still active, repeat your measurement a second time and average the values.

Remove the rubber stopper temporarily as you continue your calculations. Fan a little fresh air into the bottle. From your previous average, calculate the number of milliliters of oxygen consumed per minute and how much heat this represents. The animal produces 4.8 calories of heat for each milliliter of oxygen consumed. A calorie, as used here, is $\frac{1}{1000}$th of a kilocalorie.

When the mouse or rat becomes quiet, gently fan more air into the bottle and reinsert the stopper. If the stopper disturbs the animal, wait another minute or two, then remeasure the rate of oxygen consumption. If the animal is still quiet, take a second measurement and average them. Does the bubble descend more slowly than when the animal was active? Is the oxygen consumption less? Is the heat production less?

When finished, immediately unstopper the bottle and return the animal to its cage. If you remove the tape from the cylinder, the animal will eventually back out of its own accord.

While you measured the oxygen consumption, you might have noticed the bubble oscillating. It moves as the animal breathes—breathing that is much faster than in humans. The breathing rate parallels the higher rate of oxygen consumption and heat production per unit of weight in the small mammal.

Materials

- Two 1000-milliliter beakers or similar containers
- Thermometer
- Refrigerator or ice
- An assistant (optional)

DETERMINING HEAT LOSS FROM A HAND

Exercise of a hand or any part of the body causes a local increase in its oxygen consumption, heat production, and heat loss. Also, exercise causes dilation of blood vessels in the skin, allowing more warm blood to flow through them, causing more heat loss. Oppositely, cold exposure causes constriction of blood vessels in the skin, reducing heat loss.

In the following experiment, you will measure (1) heat loss from a hand plunged in cold water, (2) reduced heat loss from a hand previously exposed to cold, and (3) increased heat loss from a hand that is exercising. You can do this experiment alone, but the results will likely be more accurate if you have a helper.

Obtain two 1000-milliliter beakers or similar containers. Pour 500 milliliters of water into each beaker (enough to immerse your hand) and cool them in a refrigerator or with ice to 10° C (50° F).

Use the first beaker and a thermometer to measure a resting level of heat loss. Starting with the water at 10° C, place your hand in the beaker with the thermometer, keeping the tip of the thermometer away from the hand (FIG. 27-2). Stir the water frequently to distribute the heat. Record the temperature rise of the water at 1-minute intervals for 5 minutes.

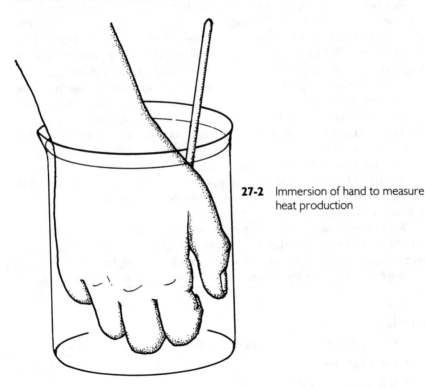

27-2 Immersion of hand to measure heat production

Remove your hand from the first beaker and immediately immerse it in the second beaker at 10° C. Again record the temperature at 1-minute intervals for 5 minutes. Because the hand is still cold and the blood vessels still constricted from previous exposure, the temperature of the water rises more slowly. Less blood flows through the constricted vessels, releasing less heat.

When you finish, notice that the hand you withdraw from the water is paler than the other hand. The paleness results from the constriction of blood vessels.

Wait an hour or longer before performing a third test on the effects of exercise. When the hand is comfortably warm, place it in a beaker of water at 10° C. Exercise the hand by moving all fingers rapidly, taking care not to spill water from the beaker. Determine the temperature rise at 1-minute intervals for 5 minutes. Exercise increases heat production and loss.

Place only a thermometer in a beaker refilled with water at 10° C. Note the very slow rise in temperature at 1-minute intervals for 5 minutes. By subtracting the temperature increase in this beaker from that in the others, you can calculate the increases resulting from the hand only.

OTHER ACTIVITIES

For comparison, place a lizard, snake, or frog in the bottle you designed for the mouse or rat. To keep the animal away from chemicals, wrap the soda lime in a layer of cloth. You need not use the wire cylinder. Instead, simply wait till the animal becomes quiet; then form a soap bubble at the end of the pipette, and time its descent.

The metabolic rates of cold-blooded animals (ectotherms) are much lower than those of warm-blooded animals (endotherms). How does this affect the rate at which the bubble traverses the tube? Does the bubble work also as an indicator of breathing rate? If so, how does this rate compare with that of the mouse or rat?

Chapter 28

Thyroids of tadpoles and mice

*I*n the neck, slung across the windpipe like saddlebags, is a two-lobed gland called the *thyroid* (FIG. 28-1). Like other *endocrine* glands, it secretes *hormones*—chemical messengers that direct bodily functions. For example, thyroid hormones cause tadpoles to develop into frogs and babies into normal adults.

When a biologist surgically removes the thyroid glands from tadpoles, they remain tadpoles. They grow bigger but never change into frogs. If the biologist later adds a thyroid extract to their water, the overgrown tadpoles become frogs. The thyroid contains factors, it seems, that cause the resorption of gills and tails and the growth of legs. The factors are *thyroxine* and other thyroid hormones.

Thyroid hormones also cause growth in humans. If the thyroid gland is missing or inactive at birth, and if replacement hormones are not given, the child becomes a slow-growing, dimwitted *cretin*. The fully grown cretin is childlike in size—sometimes less than three feet tall—misshapen in body, sluggish in movement, and infantile in behavior.

Thyroid hormones stimulate metabolism as well as growth. The word *metabolism* denotes both the synthesis and breakdown of nutrients, though it is the breakdown we are concerned with here. Thyroid hormones promote the breakdown of carbohydrates, fats, and proteins, to produce ATP, heat, and energy. Oxygen is consumed in the process. If the thyroid glands become overactive, the consumption of oxygen and production of heat increase up to 50 or 100 percent; the patient becomes hyperactive and hot.

If instead the thyroid glands are underactive, as in cretins, the metabolism decreases. There is less ATP produced, less energy evolved, less oxygen consumed, and less heat produced. Cretins, therefore, are sluggish and cold. They move and think slowly. They consume 30 to 40 percent less oxygen than normal people.

188

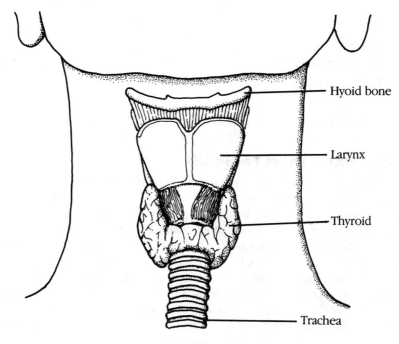

Hyoid bone

Larynx

Thyroid

Trachea

28-1 Thyroid gland

Adults can also develop thyroid deficiencies. Being grown, they do not have changes in height or mental development but do have slowness of metabolism, movements, and thought. Consider, for an extreme example, the patient of Dr. Richard Asher of London. When Asher visited the home of a sick child, he noticed an old man sitting quietly—very quietly—in a corner. The man did not move, did not respond when spoken to, and felt cold. His pulse was slow and weak. The family said the man was Uncle Toby who had "hardly moved in seven years." Because his coma developed gradually, they simply accepted it.

Dr. Asher took Uncle Toby to a hospital where his temperature registered 20° C (68° F). Suspecting a thyroid deficiency, Asher gave him thyroxine. In a few weeks, his body rewarmed, and he spoke of the events of seven years ago as if it were yesterday. When asked about his earlier feelings, he said he was "sort of cool, sort of lazy, slowed down, you know," and supposed he had "passed out . . . for a day or two."[1]

Thyroid hormones contain *iodine*. Where vegetables grow in iodine-deficient soil, people eat less iodine and make less hormones. To compensate, the thyroid glands enlarge, forming *goiters*. Being larger, the glands extract more iodine from the increased blood that flows through them. This response offsets the deficiency of iodine, usually allowing normal or near-normal production of thyroid hormones.

We now add iodine to drinking water and salt. This addition prevents thyroid deficiencies.

Materials

- Tadpoles
- Thyroid powder
- Tincture of iodine
- Thiouracil
- Four 1-gallon containers of pond water

MAKING FROGS FROM TADPOLES

Collect eight large tadpoles by seining or dipping them from a pond, or buy them from a supply house (Appendix B). For best results, choose tadpoles that are just starting to develop tiny limb buds for their hind legs.

To hold the tadpoles, fill four 1-gallon aquariums or other containers with pond water. Put two tadpoles of equal size in each of the four containers. Then add 0.2 gram of thyroid powder to the first container, 4 milliliters of tincture of iodine to the second, 0.8 gram of thiouracil to the third, and nothing to the fourth.

Feed the tadpoles with boiled lettuce or, better yet, with the moss of moss-covered rocks that you place in the containers. Change the pond water, thyroid powder, iodine, and thiouracil every few days to prevent fouling of the water.

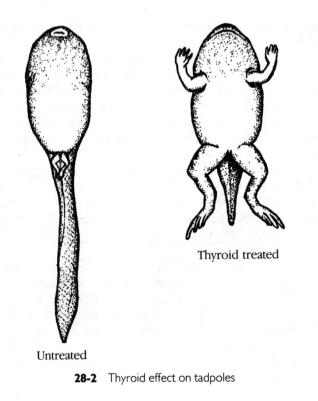

Thyroid treated

Untreated

28-2 Thyroid effect on tadpoles

If you select tadpoles whose hind-limb buds have just started to grow, you will see results in a few days. The tadpoles given thyroid powder and usually those given iodine change soon into frogs. In contrast, the untreated tadpoles change slowly and the thiouracil-treated tadpoles remain tadpoles (FIG. 28-2).

Treatment with thiouracil has the same effect as surgical removal of the thyroid glands. The tadpoles grow to a large size but never become frogs. Thiouracil acts by blocking the incorporation of iodine into thyroid hormones. There is not enough hormone produced to support metamorphosis.

When you finish the experiment, return the tadpoles and frogs to their original pond.

CHANGING THE METABOLISM OF A MOUSE

For this experiment, use a mouse or small rat and the apparatus described in chapter 27. This apparatus consists of a wire cylinder which holds the animal, a bottle containing an absorbent for carbon dioxide, and a pipette. As the animal consumes oxygen, a soap bubble moves down the pipette, indicating the rate of oxygen consumption.

CAUTION: Have an experienced person, such as the owner of a pet store, show you how to handle the mouse or rat gently and safely.

When ready, weigh the animal, direct it into the wire cylinder, and measure its oxygen consumption while it rests. Take four measurements, average them, and calculate the oxygen consumed per 100 grams of body weight for 1 minute. For example, if your mouse weighs 20 grams and consumes 0.5 milliliter of oxygen per minute, then it consumes

$$\frac{0.5 \text{ ml/min}}{20 \text{ g}} \times \frac{100 \text{ g}}{100 \text{ g}} = \frac{2.5 \text{ ml/min}}{100 \text{ g}}$$

Now mix 1 gram of thiouracil with 100 grams of ground food, either dog, rat, or mouse pellets. Feed the animal all it wants of the mixture for six weeks. Then reweigh the animal, remeasure its oxygen consumption four times, and recalculate the milliliters of oxygen consumed per minute. Does the mouse now consume less oxygen? If so, does it also produce less heat? Thiouracil blocks the incorporation of iodine into thyroid hormones, reducing the amount of hormones available. The animal, therefore, becomes hypothyroid and a bit slower than before.

Because thyroid hormones are deficient, the mouse or rat develops a goiter. Its thyroid gland now is two to four times larger than it was six weeks ago. The enlargement allows the animal to remove more iodine than otherwise from the blood. Try to feel or see the gland through the skin. It has one lobe on each side of the trachea (windpipe), as in humans, but you will probably not find it. Even the enlarged gland is tiny.

Return the mouse or rat to its normal diet. The goiter will gradually shrink and the animal will recover its original rate of oxygen consumption. Check the rate again after one or two months.

OTHER ACTIVITIES

Try the effects of thyroid hormones on aquatic animals other than tadpoles. In which, if any, do the hormones speed conversion into adults?

Also, try the effect of thyroid hormones on the metabolism of a mouse or rat, using the method just described for thiouracil. To do this, grind dog, rat, or mouse chow into powder, and to each 100 grams of this add 2 grams of thyroid powder. Measure the oxygen consumption of the animal before adding the hormone to its food and at the end of six weeks. Have the oxygen consumption and heat production increased?

Endnotes

1. Sacks, Oliver. "Rude awakening." *Discover*, Feb., 1988: 56,58.

Chapter 29

Digestion

There was a shot. Alexis St. Martin fell to the floor of the trading post in Fort Mackinac, Michigan Territory, 1822. Dr. William Beaumont came, but St. Martin would not let him suture the wound. As it healed, therefore, it left an opening from the abdomen into the stomach—a hole through which Beaumont could watch digestion.

When St. Martin's stomach was empty, said Beaumont, it was "contracted upon itself."[1] Its lining was folded, pale pink, soft as velvet, and "constantly sheathed with a mucous coat."[2] It contained little or no digestive juice.

When the stomach was filled, it stretched to hold the food, became redder with increased blood flow, and secreted gastric juice through the mucous coat. The juice formed in proportion to the amount of food eaten, and tasted "perceptibly acid."[3]

Then waves of contraction swept down the stomach. These contractions caused "a constant churning of its contents, and admixture of food and gastric juice."[4]

Beaumont tested the effects of the gastric juice on different foods. In one experiment, he drained the juice into a vial to which he added boiled beef. Then he heated the juice to the temperature of St. Martin's stomach. After 40 minutes, he saw that "digestion had commenced over the surface of the meat." In two hours, "the cellular texture seemed entirely destroyed, leaving the muscular fibers loose and unconnected, floating in fine small shreds." In 10 hours, "every part of the meat was completely digested."[5]

Typically, food remained in the stomach two to four hours— not long enough for complete digestion. The breakdown continued, however, in the small intestine, where the food passed next.

As biologists found later, the salivary glands begin digestion by secreting *amylase*, an enzyme that breaks down carbohydrates. The amylase is carried into the stomach, where it acts until gastric acid enters the food. Simultaneously, the stomach secretes *pepsin*, an enzyme that digests proteins. Then the pancreas secretes *proteinases*, *lipase*, and another amylase into the intestine to continue the digestion of proteins, fats, and carbohydrates. Finally, the intestine itself secretes enzymes that complete digestion, and it absorbs the products.

Materials

- Potato, honey, margarine, egg white, and other foods
- Unglazed paper, such as wrapping paper
- Tincture of iodine, Benedict's solution, and biuret reagent
- Beaker or pan of boiling water
- Test tubes
- Test-tube holder

SELECTION OF FOODS FOR ANALYSIS

You will be analyzing different foods for the presence of starch, sugar, fat, and protein. For your first tests, choose foods that definitely contain these ingredients. Potatoes, for example, have starch; honey has sugar; margarine has fat; and egg white has protein. Continue by testing other foods to find their ingredients.

TEST FOR STARCH

Drop some tincture of iodine on a slice of potato or other food (FIG. 29-1). If the iodine turns blue or black, the food contains starch, a carbohydrate. If the solution remains brown, the food is starchless.

TEST FOR SUGAR

Put several drops of honey or other food, and several drops of Benedict's solution into a test tube (see Appendix B for suppliers). Separately heat a beaker or pan of water until the water boils. Place the test tube with its contents in the beaker or pan, allowing time for the contents also to start boiling. If the solution turns red, the mixture contains a large amount of a reacting sugar—such as glucose, fructose, lactose, or maltose. If it turns orange, yellow, or green, it contains progressively less sugar. If it stays blue, it has no sugar. Table sugar (sucrose) does not react with Benedict's solution.

TEST FOR FAT

Place margarine or mashed food on unglazed paper, such as wrapping paper. If a translucent spot appears and remains, the food contains fat. If the spot dries and disappears, the food contains water but not fat.

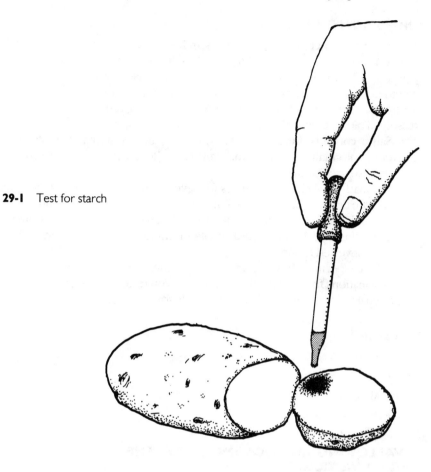

29-1   Test for starch

TEST FOR PROTEIN

CAUTION: The biuret reagent used in this test contains dilute sodium hydroxide. Use adult supervision and wear goggles. If the solution gets on your skin or clothes, wash it off with a large volume of running water.

Put several drops of biuret reagent and the same volume of egg white, milk, or mashed food in a test tube, beaker, or cup (see Appendix B for suppliers). If the blue reagent turns violet, the food contains protein or partially digested protein.

When you finish, flush the reagent down the drain with a large volume of running water.

Materials

- Unsweetened cracker
- Tincture of iodine
- Benedict's solution

DIGESTION BY SALIVA

Test an unsweetened white cracker for starch and sugar by applying tincture of iodine to one piece of it and Benedict's solution to another, as described previously. You will likely find starch, not sugar. Also test your saliva to find whether it contains starch or sugar. Then chew another cracker in your mouth for five minutes without swallowing it. When finished, test your saliva and the dissolved bits of cracker for starch and sugar. What happens?

Saliva contains amylase, an enzyme that digests starch. The starch is converted into maltose, a sugar that causes the Benedict's solution to change from blue to green or yellow.

Ordinarily, we chew substances for several seconds, then swallow them; there is little time for digestion in the mouth. But the saliva in food continues to digest starch in the stomach until the hydrochloric acid of the stomach penetrates the food. Because the acid takes minutes to penetrate, some of the starch is converted to sugar.

When experimenting, use *controls*—parallel tests—to eliminate alternative explanations for your results. This is why you tested an intact cracker and saliva separately for sugar before testing a dissolved cracker.

Materials

- Mirror
- Cup of water
- Alcohol
- Stethoscope

SWALLOWING AND MOVEMENTS OF THE DIGESTIVE TRACT

Stand in front of a mirror as you gulp water. Consciously analyze the movements of your tongue and *larynx* (Adam's apple) as you repeatedly swallow the water. Notice that the tongue moves up and back, pushing the water into the *pharynx* (throat). Then the larynx bobs up. As it bobs, it moves the *epiglottis* at the top of the larynx into a position that keeps water out of the lungs. Instead the water is squeezed downward by a ring of muscular contraction called *peristalsis*. This contraction moves the water from the pharynx to the tubular *esophagus* and from the esophagus to the stomach. The walls of the pharynx and esophagus are formed by muscle that does the squeezing.

You can time peristalsis as follows. Clean the earplugs of a stethoscope with alcohol, and put them in the openings of your ears. Drink water into your mouth without swallowing it. Lie down. Place the disk of the stethoscope on your abdomen slightly to the left of the midline near the end of your sternum (breastbone). Now swallow the water and listen. How many seconds does it take to hear a gurgle? The gurgle occurs as peristalsis squeezes the water through the *gastroesophageal sphincter* into your stomach. (The sphincter is a circular band of muscle at the end of the esophagus.) By lying down before swallowing, you have shown that peristalsis can move the water to your stomach without the help of gravity.

Now move the disk of the stethoscope toward your navel and the underlying small intestine. Listen to the intestinal movements. If you have gas bubbles among the intestinal juices and food, you will hear them being squeezed this way and that. Some of these movements are peristaltic—that is, movements that sweep the juices and partially digested food forward—but most of the movements are segmental. *Segmentation* squeezes the intestinal contents forward, then backward, mixing the juices with the food and pressing the digested nutrients nearer the wall for absorption into the body.

OTHER ACTIVITIES

A product called "pancreatin" contains all the enzymes of pancreatic juice. If you put a pinch of pancreatin in water and pour it into sausage casing or dialysis tubing along with mashed foods, you will have a digestive arrangement similar to that of the small intestine. Tie the tubing to a support that allows the filled part of the tube to droop into a beaker of water. Heat the beaker and its contents to 38° C (100.4° F). The enzymes of pancreatin digest the mashed foods, and the digested molecules diffuse from the tube into the water. Test the water periodically for sugars by using Benedict's solution, for fatty acids by using litmus (an indicator that turns red in the presence of acids), and for partially digested proteins by using biuret reagent.

Endnotes

1. Beaumont, William. *Experiments and Observations on the Gastric Juice and the Physiology of Digestion*, p. 21. Plattsburgh, New York: F. P. Allen, 1833.

2. Beaumont, p. 278.

3. Beaumont, p. 85.

4. Beaumont, p. 278.

5. Beaumont, p. 128.

Part 9

Heart and circulation

Chapter **30**

Anatomy of
a heart

"*I* almost believed," said William Harvey in 1628, "that the motion of the heart was to be understood by God alone." Its contractions took place "in the twinkling of an eye, like a flash of lightning."[1] Yet, by examining the slower heartbeats of cold-blooded animals and the direction of blood flow in arteries and veins, Harvey made a great discovery: Blood circulates.

To find how the heart works, Harvey compressed first the vein entering it, the *vena cava*, then the artery leaving it, the *aorta*, to see where blood would drain or collect. By pinching off the vena cava in a snake, he said,

> . . .you will see that the space between the finger and the heart is drained at once, the blood being emptied by the heart beat. At the same time, the heart becomes much paler even in distension, smaller from lack of blood, and beats more slowly, so that it seems to be dying. . . .
>
> On the other hand. . .if you ligate or compress the artery a little distance above the heart, you will see the. . .heart. . .become greatly distended and very turgid, of a purple or livid color, and, choked by the blood, it will seem to suffocate. . . .[2]

Holding a heart in his hand, Harvey felt its muscle contract, then saw the arteries expand. Cutting into its ventricle, the main cavity of the heart, he saw it spurt blood as it contracted. Cutting into the nearby arteries, he also saw them spurt blood as the ventricle contracted. The ventricle, he concluded, pumps blood into the arteries.

Blood circulates, he found, from the arteries to the veins and back to the heart. Otherwise, the pumping of blood by the heart would overinflate the arteries. And the veins, he saw, had one-way valves that allowed blood to pass only to the heart.

From these and other observations, Harvey concluded that "blood by the beat of the ventricle flows through the lungs and heart and is pumped to the whole body. There it passes through pores in the flesh into the veins through which it returns from the periphery everywhere to the center, from the smaller veins into the larger ones, finally coming to the vena cava" and right atrium.[3] Marcello Malpighi later identified Harvey's "pores in the flesh" as *capillaries* connecting arteries to veins.

Materials

- Dissecting tray
- Dissecting instruments
- Surgical gloves
- Beef, sheep, pig, or other mammalian heart

OUTSIDE THE HEART

From a butcher or biological supply house, get a beef, sheep, or pig heart with the major vessels attached. For fresh hearts, check the yellow pages of your phone book under "Meat." Ask a meat packer to specially cut the heart to leave its large vessels attached.

CAUTION: If instead you buy a preserved heart, rinse it thoroughly with water, and handle it with surgical gloves. If you get preservative on your skin, wash it off repeatedly with soap and running water.

Begin with the coronary arteries on the outside of the heart (FIG. 30-1). Strangely, blood within the heart does not directly nourish its own muscle, or *myocardium*. Instead the coronary arteries do. They carry nutrients and oxygen from the aorta to the myocardium.

The coronary arteries and their associated veins are often surrounded by fat, making them sometimes easier, sometimes harder to see as they cross the heart wall. Two coronary arteries, a left and right, arise from the aorta, where it first leaves the heart. These arteries run across the ventral (front) side of the heart. Thinner-walled *cardiac veins* run near the arteries, returning blood to a veinlike *coronary sinus* on the dorsal (back) side of the heart. The blood then drains into the right atrium, ready again to be circulated through the heart.

If the coronary arteries become blocked inside—as by fat or a blood clot—the result is a *coronary occlusion*. The blockage deprives the heart of nutrients and oxygen, so it cannot pump effectively. Coronary occlusion is the leading killer of people in middle or old age.

Now trace the course of blood through the heart, starting with the thin-walled venae cavae (FIG. 30-2). Do not yet cut open the heart; it will be easier to see structural relations by remaining outside. Blood returning from the head, neck, and arms enters the *superior vena cava*, passing from there into the *right atrium*. Likewise, blood from the legs and trunk enter the *inferior vena cava*, passing into the right atrium. Run a finger or probe through the superior and inferior venae cavae to feel their connections with the atrium. Blood is drawn from the right atrium into the *right ventricle* as the ventricle

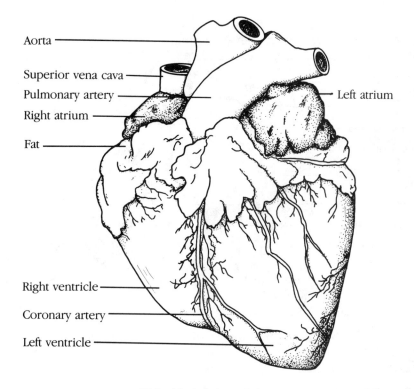

Aorta

Superior vena cava

Pulmonary artery

Right atrium

Fat

Left atrium

Right ventricle

Coronary artery

Left ventricle

30-1 Ventral view of a heart

relaxes and expands between contractions. Expansion of the ventricle creates suction. Then the atrium contracts, squeezing a little more blood into the ventricle before it contracts. Contraction of the right ventricle pumps blood through a *pulmonary artery* and its branches to the lungs. Here the blood receives oxygen and releases carbon dioxide. The oxygenated blood circulates through *pulmonary veins* to the *left atrium*, and from there to the *left ventricle*.

Press your fingers against the two atria. Notice how thin-walled they are. They need little muscle to pump blood into the adjoining ventricles.

Now press your fingers against the walls of the two ventricles. Notice that these walls are thicker than those of the atria, and that the left ventricular wall is thicker than the right. The left ventricle is more muscular because it pumps blood through the aorta to the entire body except the lungs. The right ventricle is less muscular because it pumps blood only to the nearby lungs.

Now compare the wall of the aorta with that of the pulmonary artery. The left ventricle generates a much higher blood pressure in the aorta than the right ventricle generates in the pulmonary artery. Consequently, the wall of the aorta is more muscular than that of the pulmonary artery.

Also compare the walls of the aorta and pulmonary artery with those of the venae cavae. The blood pressure in the venae cavae is approximately zero where blood enters the heart. The caval walls, therefore, are extremely thin.

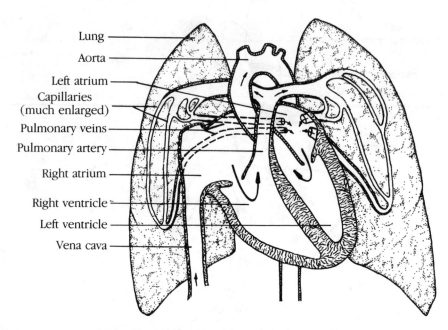

Lung

Aorta

Left atrium

Capillaries
(much enlarged)

Pulmonary veins

Pulmonary artery

Right atrium

Right ventricle

Left ventricle

Vena cava

30-2   Circulation of blood through the heart and lungs

INSIDE THE HEART

Open the heart by slitting first the right atrium and right ventricle. Notice the beauty and strength of their interwoven, reinforced fibers (FIG. 30-3). Notice also the muscular walls, heavier in the ventricle. Feel the openings of the superior and inferior venae cavae into the atrium. See the right *atrioventricular valve* (AV valve) that directs blood in one direction only, from the atrium to the ventricle. This valve is also called the tricuspid valve, because it has three flaps or cusps forming the valve. The cusps might have been cut and partly destroyed in opening the heart.

Pull the string-like *tendinous cords* attached to the valve. The ventricle draws these cords downward when it relaxes, pulling the flaps of the AV valve open. This opening allows atrial blood to flow into the ventricle. Conversely, the ventricle pushes the cords upward when it contracts, allowing the flaps of the AV valve to close. Simultaneously, the contracting right ventricle forces blood upward through the *pulmonary semilunar valves* into the pulmonary artery. The word semilunar means "half moon." There are three of these half-moon, one-way valves that work as a unit, opening when the ventricle contracts, closing when it relaxes.

Now slit open the left atrium and left ventricle. The structure and operation of the left side of the heart are similar to the right, but the muscular wall of the left ventricle is extremely thick. It is thick because it pumps blood through the aorta to distant parts of the body. Find the four pulmonary veins where they enter the left atrium. They carry oxygenated blood from the lungs to the heart. Locate the left atrioventricular valve, sometimes called the mitral

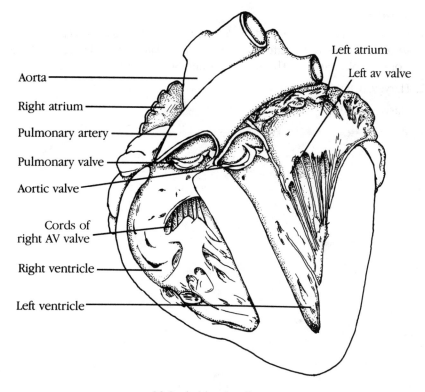

Aorta

Right atrium

Pulmonary artery

Pulmonary valve

Aortic valve

Cords of
right AV valve

Right ventricle

Left ventricle

Left atrium

Left av valve

30-3   Inside a beef heart

or bicuspid valve. It has only two flaps, not three as does the right valve. Finally, locate the *aortic semilunar valves* that direct blood from the left ventricle to the aorta.

OTHER ACTIVITIES

Compare other vertebrate hearts with those of mammals. Fish hearts have only two chambers—an atrium and a ventricle. Amphibian and reptilian hearts have three chambers—two atria and one ventricle. Bird hearts have four chambers like those of mammals.

Also trace blood vessels from the heart to distant organs. Dealers often inject the vessels of their preserved specimens with colored latex rubber, a valuable aid for the beginner. Arteries are red, veins are blue.

You might also get animals with dyed lymphatic vessels. These vessels are narrow and thin-walled, hard to see, but the dye reveals them. They pass through lymph nodes in which phagocytes destroy invading microorganisms, such as bacteria. Have you noticed that the lymph nodes in your own body become swollen during infections? When they are swollen, the phagocytes are actively destroying the microorganisms that cause the infection.

Endnotes

1. Harvey, William. *Anatomical Studies on the Motion of the Heart and Blood in Animals*. 1628. Frankfurt, Germany. Translated by C. D. Leake. 1970. Courtesy of Charles C. Thomas, Publisher, Springfield, Illinois, p. 25.

2. Harvey, p. 84.

3. Harvey, p. 104.

Chapter **31**

Heartbeats
and pulses

*I*f you listen to the heart as a person starts exercising, you will hear a progressively louder and faster beat. Similarly, if you place your hand over the heart, you will feel a stronger and faster beat. Nerves release adrenaline into the heart during exercise, causing it to beat harder and faster, pumping more blood from the heart through the arteries. Loaded with oxygen, the blood rushes to hard-working muscles.

When people think of exercise or become emotional, their bodies prepare. The *thought or emotion* signals the cardiovascular center in the brain's medulla (FIG. 31-1). This center connects with sympathetic nerves to the heart, the ones that increase its rate and output. Anyone who doubts the mind's influence on the heart should consider the records of a balloon ascension in 1956. The balloon drifted into a hazardous thunderstorm. Telemetered data from the confined crew showed heart rates over 170 beats per minute! Compare this with the normal rate of 70 to 80 beats per minute. Similarly, the heart rates of trained sprinters increase to 150 beats per minute in the final seconds before a race.

The start of exercise causes other signals from the brain's medulla that increase the heart's output. Because this increase occurs after exercise begins, it is not due to anticipation. Instead it comes from nerve signals starting in *stretch receptors of muscles, tendons, and joints*. These signals pass to the cardiovascular center of the medulla that speeds the heartbeat. In addition, the *motor area* of the cerebrum directs excitatory signals through the medulla to the heart at the same time it directs excitatory signals to contracting muscles.

As exercise begins, muscles consume more oxygen and produce more carbon dioxide. Consequently, the levels of these gases change in the blood. But the changes in oxygen and carbon dioxide have no effect on the heart until the gases have time to circulate to the aorta in the chest and to the carotid

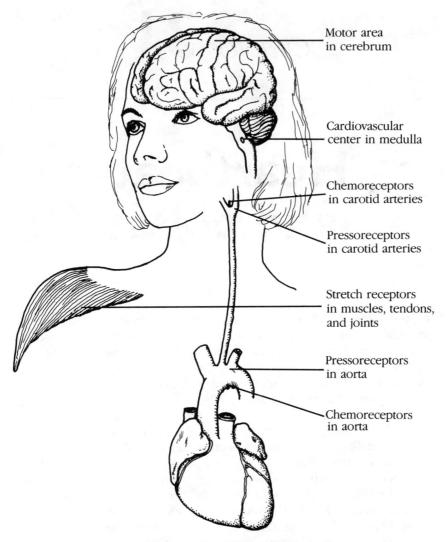

Motor area
in cerebrum

Cardiovascular
center in medulla

Chemoreceptors
in carotid arteries

Pressoreceptors
in carotid arteries

Stretch receptors
in muscles, tendons,
and joints

Pressoreceptors
in aorta

Chemoreceptors
in aorta

31-1 Neural and chemical control of the heart

arteries in the neck. These vessels contain *chemoreceptors* sensitive to changes in the levels of the gases. The chemoreceptors act through the medulla to cause a slight further increase in the rate and force of the heartbeat.

If the blood pressure were to rise too high, some of the vessels might burst from the excessive pressure. This occasionally happens, causing a stroke. But it usually does not happen, because the circulatory system contains safety gauges. The gauges—called *pressoreceptors*—are stretch-sensitive nerve endings in the aorta and carotid arteries. High blood pressure expands the vessels, stretching the receptors. In response, the receptors signal the medulla to slow the heartbeat. The slowing reduces the blood pressure. Conversely, if the pressure becomes too low, the pressoreceptors signal the medulla to speed the heartbeat. The speeding increases the blood pressure. The role of the pressoreceptors, therefore, is to keep the blood pressure stable.

Materials

- Human subject
- Stethoscope (or funnel, Y-tube, and rubber tubing)
- Isopropyl alcohol
- Watch or clock with a second hand

LISTENING TO THE HEARTBEAT

In earlier times, physicians placed their hands or ears on a patient's chest to feel or hear the heartbeat. But in 1816 a French doctor, René Laennec, was faced with an embarrassing problem. He was unable to feel the heartbeat of a young woman excessively padded with fat and clothing. Rather than use his ear directly, he decided to roll paper into a cylinder and listen through this. It worked, and from this meager beginning, our present-day *stethoscopes* evolved.

You can purchase or make a stethoscope similar to your doctor's. Get a funnel, a Y-tube, and some rubber tubing to fit them. Cut the tubing to provide two pieces about 40 centimeters (16 inches) long and one piece about 20 centimeters long. Fasten the long rubber tubes to the branches of the Y-tube to use as earpieces. Fasten the short rubber tube from the free end of the Y-tube to the funnel (FIG. 31-2).

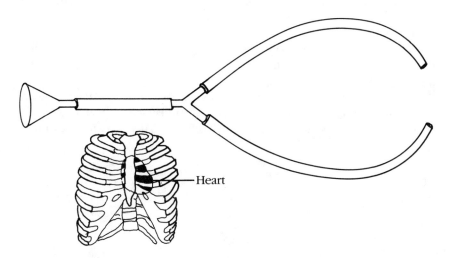

31-2   Homemade stethoscope and location of the heart

Clean the earpieces of your stethoscope with alcohol, and let them dry. Then place the earpieces in your ears. Your ear canals angle slightly forward in your skull, so you should angle the earpieces accordingly. Now hold the disk or funnel end of the stethoscope against the chest of a human subject, just to the left of the sternum (breastbone). This is where the heart is. Explore until you hear strong sounds, usually toward the lower level of the sternum (breastbone).

Listen for a double sound, lub-dup, then a pause of less than one second, and another lub-dup. The *lub* sound is louder and longer, more booming, than the dup sound. The lub results from closure of the *atrioventricular valves*, the valves that separate the atria from the ventricles (chapter 30). These valves shut at the start of ventricular contraction. Shortly after, the shorter, higher-pitched *dup* sound results from closure of the *aortic and pulmonary valves*. These valves shut at the start of ventricular relaxation.

Abnormal heart sounds are called *murmurs*. In some people, for example, the blood swishes as it flows backward through valves that cannot fully shut. In other people, the blood screeches as it flows forward through valves that are too narrow. If you know someone with a heart murmur, ask if he or she would mind you listening to it.

Materials

- Human subjects
- Watch or clock with a second hand

FEELING AND COUNTING THE PULSE

You can feel the pulse in several places. Among these are the carotid arteries of the neck (alongside the trachea), the temporal arteries at the sides of the head, and the *radial arteries* of the wrists. To detect a radial pulse, place your thumb on the back of a wrist and your fingers over the radial artery on the underside of the wrist (FIG. 31-3). Exert only a medium pressure with your fingers to feel the pulse. Count the rate of arterial expansions per minute. This count is called the pulse rate.

As you feel the pulse of different people, notice how regular or irregular it is and how strong or weak its amplitude. If the pulse skips a beat occasionally or if the pulse is unusually weak, making it difficult to detect, something might be wrong.

Materials

- Human subject
- Stethoscope
- Watch or clock with a second hand

COMPARING THE HEARTBEAT WITH THE PULSE

Contraction of the left ventricle generates the pulse. The contraction squeezes blood into the aorta, expanding its wall and the walls of all arteries that follow

31-3   Taking the pulse

it. The pulse wave, which resembles the wave passing down a violin string plucked at one end, travels faster than the flow of blood. It moves from the heart to the wrist in about ½ second.

Listen to the heartbeat of a person at the same time you feel his or her radial pulse. Each lub of the lub-dup results from the contraction of the ventricles. Is the heart rate per minute the same as the pulse rate per minute? In other words, is the frequency of the lub sounds the same as the frequency of the pulse waves? They should be. You will note, however, that the lub sounds come about ½ second before the pulse waves. This is the time taken for the pulse waves to travel from the heart to the wrist.

Materials

- Human subjects who have not recently exercised

EFFECT OF POSITION ON PULSE

Get one or more subjects who have not been exercising in the last half hour. Have them lie down for a few minutes, then measure their pulse rates. Do this as they recline, then sit, and finally stand. Most subjects have slight increases in pulse rate when they sit and stand.

The pressoreceptors—stretch-sensitive receptors in arteries—are largely responsible for these increases in pulse rate. When the subjects sit and stand, gravity pulls blood from their heads to their feet, lowering the blood pressure in the upper halves of their bodies. The pressoreceptors then signal their medullas to speed their heartbeats. The faster heartbeats return the blood pressure to the normal level.

Materials

- Human subject
- Watch or clock with a second hand
- An exercise bicycle (optional)

EFFECT OF EXERCISE ON PULSE

Any type of strenuous exercise will suffice for this experiment. Bicycling, running, and gymnastics, for example, all cause increases in heart and pulse rate. If you have access, however, to an exercise bicycle, use it because it allows you to measure the pulse rate both during and after the ride. Indeed, some bicycles have built-in electronic devices that can be clipped to the earlobes to measure the pulse.

First seat your subject on the bicycle and measure his or her resting pulse rate. Then have the subject pedal vigorously at a constant rate—let us say 20 kilometers per hour (12 miles per hour)—for 10 minutes. Measure the pulse rate every minute during the first 4 minutes of pedaling and every 2 minutes after this. The changes in pulse rate will be more rapid at first.

For convenience, you can measure each pulse rate for only 30 seconds and double your readings. For example, if you measure 50 pulse beats in 30 seconds, multiply this by two to get the rate of 100 beats per minute.

As you measure the pulse, note that it becomes stronger during exercise than before. It is stronger because the heart contracts more forcefully, pumping more blood into the arteries.

At the end of 10 minutes, have the subject stop pedaling. Immediately measure the pulse rate and continue to do so at 1-minute intervals for several minutes. Then measure the rate at 2-minute intervals until the subject recovers—that is, until the pulse is the same as before bicycling.

Recall that the heart's output increases for many reasons during exercise. These reasons include increased emotion, signals from the motor cortex to the heart, excitation of stretch receptors in contracting muscles, and excitation of chemoreceptors in the carotid arteries and aorta. Which of these factors continue to operate during recovery from exercise?

Materials

- Human subject
- Swimming pool or bowl of water

EFFECT OF SUBMERSION ON PULSE

Measure your pulse rate as you float underwater or as you submerge your face in a bowl of water. Compare this rate with your normal rate when your face or body is not immersed.

Many people experience a modest slowing of the heartbeat and pulse rate when they submerge, and some, a drastic slowing. One of my students had a progressive decrease in heartbeat from his normal 60 beats per minute to 25. Such changes occur more prominently yet in seals, ducks, and other underwater swimmers, the rate dropping to 15 or fewer beats per minute when they submerge. The demand for blood decreases during submersion because the vessels in their limbs constrict.

Materials

- Human subject
- Step or bench that measures 20 inches high for men or 17 inches for women
- Watch or clock with a second hand
- A metronome (optional)

STEP TEST FOR PHYSICAL FITNESS

In this test the subject steps up and down on a step or bench that is 20 inches high for men and 17 inches high for women. Each step is made at ½-second intervals. You call the cadence or set a metronome ticking every ½ second. Call the steps as—*up*, two, three, four; *up*, two, three, four; and so on—giving the signal "up" at 2-second intervals.

Now have the subject stand in front of the step or bench. At the command "up," have him or her place the right foot on the step; at "two," draw up the left foot and stand erect; at "three," lower the right foot to the floor; and at "four," bring down the left foot. The subject is now in the starting position again but will continue stepping as you continue calling changes in position.

Exercise the subject for 5 minutes or until he or she is prematurely exhausted or is unable to maintain the stepping pace for 15 seconds. When finished, ask the subject to sit down. Jot down the duration of exercise.

Count and record a 30-second pulse rate at exactly 1 minute, 2 minutes, and 3 minutes after the exercise. Total these 1 to 1½, 2 to 2½, and 3 to 3½ minute pulse counts as illustrated below to get the subject's score.

$$\text{Fitness Index} = \frac{\text{Duration of exercise in seconds} \times 100}{2 \times \text{Sum of the three 30-second pulse counts}}$$

A score below 55 is considered poor, typical of those who are physically less active. A score of 55 to 90 is average, typical of the 2000 college students and others who were originally tested. An index above 90 is excellent and expected for long-distance runners, cross-country skiers, swimmers, and similar athletes.

Endurance and speed of recovery are the two factors used here to estimate physical fitness. During the step test, the heart rate accelerates quickly in both athletes and nonathletes. But after the exercise, recovery comes sooner in athletes. Their hearts pump more blood to their muscles.

Materials

- Pet dog
- Ball
- Watch or clock with a second hand
- Human subject
- A helper with ice cubes

HEARTBEAT DURING EXCITEMENT

I have a pet retriever, Ginger, who loves to chase tennis balls. Sometimes, while she is standing at the glass door facing our yard, I reach under her chest to feel her heartbeat. Then I say, "Ginger, I'm going to throw a ball." Immediately, her heartbeat accelerates as she anticipates the result. Similarly, her muscles tense in the chest and abdomen. As I open the door, her heart speeds further. When I throw the ball, her heart probably speeds further yet, but by then Ginger is racing across the yard.

If you have a pet dog, feel the heartbeat as you prepare to throw a ball. It will probably accelerate greatly.

For a similar experiment, measure the pulse rate of a human while an accomplice drops several ice cubes at once down the subject's back. The surprise will produce an increase in heart rate.

Materials

- Daphnia (water flea)
- Microscope
- Slide
- Bristle
- Cover glass
- A source of heat and cold

HEARTBEAT OF DAPHNIA

You can obtain *Daphnia*, the water flea, from a biological supplier or by dipping from a pond. For dipping, use a finely woven dip net. Sweep it through the water near aquatic plants.

Place several Daphnia in water on a slide, and cover them with a cover glass. Support the glass with a bristle to avoid crushing the Daphnia. Under low power, focus your microscope on one of the animals to see its rapidly beating heart (FIG. 31-4). Observe the rate of the beat when the animal is quiet and immediately after it moves. Also, warm or cool the slide to see how these changes affect the heartbeat.

Human heartbeats speed during fevers. The sinoatrial node—the pace-maker at the entrance to the heart—becomes more permeable to sodium ions from the fluid that surrounds it. The faster entry of sodium ions to the

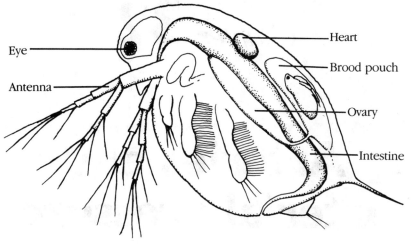

Eye

Antenna

Heart

Brood pouch

Ovary

Intestine

31-4   Heart of Daphnia

pacemaker speeds the heartbeats. Conversely, the heartbeats slow during hypothermia—that is, when the body becomes extremely chilled from prolonged exposure to cold. Sodium ions then enter the pacemaker more slowly. Poorly insulated hikers sometimes develop hypothermia during rainstorms or snowstorms.

OTHER ACTIVITIES

If you have friends with differing levels of physical fitness, ask several of them to do the bicycle ride or step test described earlier. How do their pulse rates and recovery from exercise compare? Additionally, if you have friends who are joggers and others who lift weights, compare their pulse rates during and after exercise. Which group recovers more quickly?

Two of the author's students ran alongside the Olympic distance runner Fred Wilt, and the class compared their recoveries. In a ½-mile, 3-minute run, a pace that exhausted the students, Wilt was no more fatigued than an ordinary man after a walk. After 2 minutes of recovery, his heart beat 72 times per minute, and in 20 minutes it had returned to the pre-race level of 60. The students had pulses over 120 beats per minute after the same 2 minutes, and over 100 when the athlete was normal. Blood pressure and breathing recovered similarly.

If we stay physically fit, our bodies respond effectively to exercise. Repeated exercise strengthens heart muscle as it strengthens skeletal muscle. For this reason, an athlete's heart can pump more blood during each contraction. The more forceful beat allows longer exercise and faster recovery.

Chapter *32*

Arteries and arterial pressure

Stephen Hales, an Englishman of the 1700s, thought that blood must have pressure to force it along vessels; and if it had pressure, he could measure it by inserting tubes into arteries and veins. The pressure would force the blood a measurable distance up the tubes. So he tried the experiment on a horse, getting the first-ever reading of blood pressure.

> I laid a common Field Gate on the Ground with some Straw upon it, on which a white Mare was cast on her Side, and in that Posture bound fast to the Gate. . . . Then laying bare the left Carotid Artery, I fixed to it towards the heart the Brass Pipe, and to that the Wind-pipe of a Goose; to the other end of which a Glass Tube was fixed. . . . The Blood rose in the Tube...to nine Feet six Inches Height.[1]

Biologists still use Hales' method at times to measure the blood pressure of experimental animals; but to avoid the use of ceiling-high tubes, they substitute mercury for blood in shorter tubes. Since the mercury is about 13 times heavier than blood, the pressure of blood raises the mercury only a short distance. This allows measurements in millimeters of mercury (mm Hg) rather than feet of blood. Also, electronic gauges, sensitive to pressure, can be substituted for mercury; but the gauges must first be calibrated by comparing their responses to pressure with those of mercury.

For most measurements of blood pressure, however, physicians and experimenters alike use *sphygmomanometers*. A sphygmomanometer has two parts: a tube of mercury (or an air gauge) and a cuff wrapped around the arm (FIG. 32-1). The operator pumps air into an inflatable rubber bag inside the cuff, until the cuff stops blood flow in the brachial artery of the arm. Then air is released from the bag until the artery reopens. Using a stethoscope, the operator hears the squirt of blood as the artery opens. Upon first hearing the sound, he or she notes the height of the mercury column and records it as the

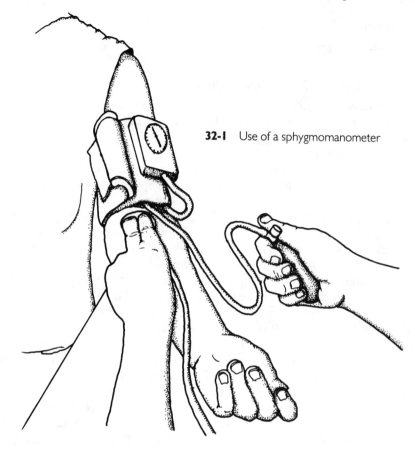

32-1   Use of a sphygmomanometer

systolic pressure, the pressure at which the contracting ventricle of the heart can barely force blood through the occluded artery. As the operator continues to deflate the cuff, the sounds get louder, then softer, and finally disappear. Upon hearing the last sound, he or she records the height of the mercury column as the *diastolic pressure*, the pressure at which the arterial walls remain continuously open.

When young adults sit, their systolic blood pressure averages 120 mm Hg. This means that the contraction, or *systole*, of their ventricles pushes arterial blood forward with a force of 120 millimeters of mercury, a force sufficient to raise a column of mercury to a height of 120 millimeters. In contrast, their diastolic blood pressure averages 75 mm Hg. This means that the relaxation, or *diastole*, of their ventricles lowers the arterial pressure to 75 millimeters of mercury; that is, it allows the column of mercury to drop to a height of 75 millimeters. The two readings are written as 120/75.

Materials

- Human subject
- Sphygmomanometer

- Isopropyl alcohol
- Stethoscope

MEASURING BLOOD PRESSURE

If possible, get help from a physician, nurse, or other person familiar with the sphygmomanometer you are using. There are several kinds of sphygmomanometers and cuffs, each slightly different in operation. The procedure that follows will work for most.

Have your subject sit at a table and rest his or her arm upon it. Fasten the sphygmomanometer cuff snugly around the bare upper arm, with the bag over the biceps and one inch above the bend of the arm (FIG. 32-1).

Clean the earplugs of the stethoscope with alcohol. When dry, insert the earplugs in your ears, angling the plugs slightly forward to fit the angle of your ear canals. Press the disk or bell of the stethoscope over the brachial artery at the bend of the arm. Screw the circular, metal valve clockwise to shut it (if the bulb of your sphygmomanometer has this kind of valve), and pump the cuff full of air by squeezing the bulb several times. As you squeeze, the pressure registers on an air gauge or mercury manometer. Stop pumping at 150 to 170 mm Hg. If your subject is normal, this pressure will occlude the brachial artery. Thus, as you listen through the stethoscope, you will hear no sound. Twist the screw to deflate the cuff, lowering its pressure slowly at 2 to 3 mm Hg per second. Some cuffs deflate automatically at this rate. As the pressure decreases, note the first, faint tapping sound, and record this figure as the *systolic blood pressure*. Each tap occurs because the ventricular contraction (or systole) forces arterial walls apart to let through a brief, noise-making spurt of blood. Continue deflation. The sound gets louder, fainter, duller, and then disappears. Blood now flows through in a continuous stream; the walls no longer collapse against each other. Record the *diastolic pressure*, the pressure at which the sound disappears. Then release air from the cuff at once.

As you measure arterial pressure, the gauge needle or mercury column oscillates in the approximate range of pressures between systolic and diastolic. These oscillations of one to two millimeters result from the change in pressure accompanying each pulse wave. The first oscillations occur just before you hear the tapping sounds that represent systolic pressure, warning you to listen carefully.

The arterial blood pressure for relaxed young men averages 120 mm systolic and 75 diastolic (written 120/75); that of women is a few millimeters lower. The normal range for systolic pressure is 90 to 140 mm.

As people age, lipids and calcium usually deposit in their vessel walls. The deposits hinder blood flow as their hearts keep pumping, causing their blood pressures to increase. We see a similar effect in the water of a garden hose when we twist the nozzle partly closed. The water pressure in the hose increases.

Materials

- Human subject
- Sphygmomanometer

FEELING THE PULSE WHILE VARYING THE CUFF PRESSURE

Keep the sphygmomanometer cuff on the subject in whom you just measured the blood pressure. Feel the pulse in the wrist of this subject, as described in chapter 31. While feeling the pulse with one hand, pump up the sphygmomanometer cuff with the other. What happens to the pulse when you exceed the level of the systolic blood pressure? It should disappear because the cuff blocks the pulse wave from the arm to the wrist. Now let the pressure slowly out of the cuff. The pulse should reappear when the pressure drops below the systolic level, as blood spurts again through the arteries.

Materials

- Human subject
- Sphygmomanometer
- Stethoscope
- An exercise bicycle (optional)

EFFECT OF EXERCISE ON BLOOD PRESSURE

This experiment resembles that of chapter 31 in which you measured the effect of exercise on the pulse. Use an exercise bicycle, if possible, because this will allow you to measure the blood pressure both during and after the exercise. Otherwise, use another exercise, such as running in place, and measure the blood pressure only in the recovery period afterward.

Obtain the blood pressure of your subject as he or she sits first in a chair and second on the bicycle. Then have the subject pedal vigorously at 20 kilometers per hour (12 miles per hour). Measure the blood pressure after 3, 6, and 9 minutes, as the subject continues pedaling. At 12 minutes, have the subject get off the bicycle and sit again in the chair. *Immediately* measure the blood pressure, and continue to remeasure it at 3-minute intervals until the subject recovers.

Exercise raises the blood pressure because the heart beats faster and harder. Thus it pumps more blood into the arteries. The pumping of blood into arteries raises the blood pressure just as the pumping of air into tires raises the air pressure.

Materials

- Human subject
- Sphygmomanometer

- Stethoscope
- 1000-milliliter beaker or similar container
- Ice
- Water
- A laboratory thermometer (optional)

EFFECT OF COLD ON BLOOD PRESSURE

Fill a 1000-milliliter beaker or similar container with about 500 milliliters of ice and water. Allow the temperature of the water to drop to 5° C (41° F) or less. Leave some ice in the water to continue melting during the experiment.

Have your subject rest for a few minutes, then measure his or her blood pressure and pulse rate. Now have the subject fully immerse his or her free hand in the ice water. Remeasure the blood pressure and pulse rate at two and five minutes after immersion. Do you get an increase in pressure and rate?

The iced water excites cold and pain receptors in the subject's hand. These receptors send signals to the brain, including the cardiovascular center in the medulla, reporting the discomfort. The medulla directs sympathetic nerve impulses to the heart, causing stronger, faster heartbeats, and to the arteries and their branches in the skin, causing them to constrict. The increased pumping of blood and constriction of arteries cause an increase in blood pressure.

When exposed to cold of this intensity, many people have an increase in both systolic and diastolic pressures of about 20 mm Hg—but the response varies greatly. You might want to check the response of several people to see the difference.

OTHER ACTIVITIES

In the hand-immersion experiment, test several subjects until you find one who has little change in blood pressure and another who has an increase in systolic pressure of 30 mm or more. Then test the parents of your subjects in the same way. The response to cold tends to be inherited.

Also compare the recovery of two runners in a fast-paced race. For your runners, choose one person who exercises infrequently and another who runs long distances several times weekly. Have the person who exercises infrequently set a steady pace for both, as fast as possible, for a run of 300 meters (0.15 mile) or more. You and others measure the blood pressures, pulse rates, and respiratory rates of the runners before the race and at 3-minute intervals after.

Endnotes

1. Hales, Stephen. *Statical Essays: Containing Haemostaticks; or, An Account of some Hydraulick and Hydrostatical Experiments Made on the Blood-Vessels of Animals*, pp. 13-17. London: W. Innys, R. Manby, and T. Woodward, 1733.

Chapter 33

Control of blood pressure

As a pilot pulls out of a steep dive, centrifugal force slams him or her into the seat and drains blood from the brain. The pilot blacks out unless blood is restored. To compensate, the medulla of the brain directs the heart to accelerate up to 150 or 200 beats per minute. The heart then pumps extra blood to the brain, so the pilot can remain conscious.

In other maneuvers, the pilot might have blood driven upward to the neck and head. To prevent the excessive pressure of blood from damaging the brain, the medulla directs the heartbeat to slow or momentarily stop.

What alerts the brain to these hazardous changes in blood pressure? We each have sensory receptors in the chest and neck that respond to the stretch or lack of stretch of arteries. These are called *pressoreceptors*. Located mainly in the aorta and the carotid arteries of the neck, the pressoreceptors help stabilize blood flow, especially to the brain. They keep the pressure high enough to prevent fainting and low enough to prevent the rupture of blood vessels (FIG. 31-1).

When the blood pressure decreases in the chest and neck, as in the pilot who pulls up from a dive, the aorta and carotid arteries partially collapse. The pressoreceptors in these arteries are less distended than before. Consequently, they send fewer impulses to the medulla of the brain, alerting it to the decreased pressure in the arteries. The medulla responds by sending impulses through sympathetic nerves to the heart and blood vessels, impulses that release adrenaline. The adrenaline causes the heartbeat to strengthen and accelerate and many blood vessels to constrict (become narrow). This combination increases the blood pressure, compensating for the deficiency.

In contrast, when the blood pressure increases greatly in the chest and neck, as in a pilot who pulls over a loop, the aorta and carotid arteries distend with blood. Distension stretches the pressoreceptors, causing them to send more impulses than usual to the medulla. The medulla responds by sending impulses through parasympathetic nerves to the heart and blood vessels, impulses that release acetylcholine. The acetylcholine causes the heartbeat to slow and weaken and the vessels to dilate, lowering the blood pressure.

The pressoreceptors benefit earthlings as well as pilots. Every time we stand, the receptors register a drop in blood pressure. When we rise from a chair or bed, gravity pulls about 500 milliliters of blood from the head and chest into the legs and feet. The heart suddenly pumps less blood. Unless the pressoreceptors respond, we might faint. They do respond, and by now you probably can predict the corrective actions of the medulla on the heart and vessels. Why do people remain conscious when they stand?

If we had no pressoreceptors, we could not respond rapidly enough to protect ourselves from moment to moment changes in arterial pressure. These changes occur often from standing and other causes, such as exercise, pain, and emotions.

Materials

- Human subject
- Sphygmomanometer
- Isopropyl alcohol
- Stethoscope
- Watch or clock with a second hand

EFFECT OF POSTURE ON PULSE AND BLOOD PRESSURE

Refer to chapters 31 and 32 where I give the procedures for measuring the pulse and blood pressure. If you do not have a sphygmomanometer to measure the blood pressure, measure only the pulse, and assume that the pressure will remain approximately constant.

Use a subject who has not recently exercised and who lies quietly for an additional five minutes. Then measure and record the blood pressure and pulse rate while the subject (1) continues to recline and (2) then stands. Has the pressure stayed about the same? What is the value of this constancy, particularly for the brain? How much, if any, did the pulse rate increase?

In most people, the heartbeat and pulse rate increase 10 to 30 beats per minute upon standing, and many of the blood vessels constrict. Recall that standing shifts blood downward to the legs and feet, lowering the blood pressure in the aorta and carotid arteries. How does this affect the pressore-ceptors, the medulla, and the release of adrenaline by sympathetic nerves? What does the adrenaline do to the heart and blood vessels to restore normal blood pressure?

Materials

- Human subject
- A board that is about 2 feet wide and 6 feet long

EFFECT OF A HEAD-DOWN POSITION ON PULSE

My students built a tilt-table to which we strapped our subjects. We could then rotate them horizontally or vertically, head up or head down. These changes markedly affected the pressoreceptors and their responses.

You can make a similar but less elaborate tilt-table by placing a long, wide board against a firmly placed couch or chair, so that the angle of the board is about 30 degrees. Using a well-rested subject, measure and record the pulse rates as he or she reclines horizontally and then stands. Next have the subject lie on the board, head downward at an angle of about 30 degrees. After the subject has rested in this position for 2 or 3 minutes, remeasure and record the pulse rate. Is it slower than in the other positions?

In the head-down position, gravity draws more blood than usual into the aorta and carotid arteries. They swell, stretching the pressoreceptors. What effect does this have on the medulla and the heart? Do your findings confirm this? Were your findings similar to those of our subject turned head up and down (FIG. 33-1)?

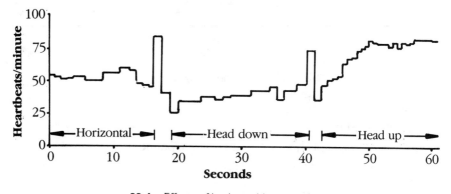

33-1　Effects of body position on pulse.

Materials

- Trumpet or other wind instrument
- Someone to play it

EFFECT OF TRUMPET PLAYING ON PULSE

Trumpeters sometimes become dizzy when playing loud, high-pitched, long-held notes. The dizziness results from the pressure in the chest when they bear down. This pressure dams back the venous blood, interfering with the flow of blood through the heart and lowering blood pressure to the brain. The reduced blood pressure acts through the pressoreceptors and medulla to alter the pulse.

Find a person who plays the trumpet or some other wind instrument, who is willing to be your subject. Ask this person to play a high-pitched note loudly and to continue playing it for a long time. Check the pulse rate and amplitude before and during the note. Also observe the swelling of veins in the head and neck and the flushing of the face. To what do you attribute the venous engorgement, the flushing of the face, the eventual weakening of the pulse, and the increase in pulse rate.

Now ask the subject to play a tune loudly as you watch the face and feel the pulse. Do you notice any changes? For comparison, see FIG. 33-2 for the effects of trumpet playing on one of my subjects.

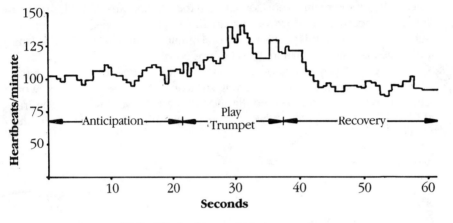

33-2   Effects of trumpet playing on pulse.

Materials

- Human subject
- Heavy weight of about 45 kilograms (100 pounds)
- Watch or clock with a second hand

EFFECT OF WEIGHT LIFTING ON PULSE

Measure the resting pulse rate of a healthy subject, then have him or her lift a weight of 40 to 50 kilograms or, alternatively, attempt to lift a weight that cannot be moved. Watch the veins in the subject's neck and head to see if they engorge, and the face to see if it flushes. Also check and record the pulse amplitude and rate.

When lifting or attempting to lift very heavy weights, subjects often hold their breath. This allows them better to anchor the muscles that do the lifting. The breath holding compresses veins in the chest, reducing the venous input of blood to the heart. It causes veins elsewhere to engorge, the face to flush, the heart eventually to pump less blood, the arterial pressure to decrease, the pressoreceptors to be less stretched, and the heart to speed.

OTHER ACTIVITIES

Measure the blood pressures and pulse rates of musicians who play different wind instruments forcefully and of singers who sing loud, long-held notes. Relate your findings to the pressoreceptors and their effects.

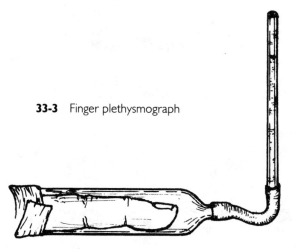

33-3 Finger plethysmograph

Construct a plethysmograph to measure changes in the blood volume of a finger. Such changes occur every time a pulse of blood enters the finger. The pulse changes as people play wind instruments, raise and lower their hands, have ice cubes dropped down their backs, and so on. Make the plethysmograph by heating polyethylene tubing and pulling it into the shape shown in FIG. 33-3. Use adhesive tape to close the junction of the tube with the finger. Fill the tube with water, excluding all bubbles, and connect it to a graduated pipet with which to measure the changes in volume.

When blood flow decreases in the finger, the volume of blood decreases and the level of water falls in the plethysmograph. Excitement or pain, for example, causes this. They release adrenaline, causing a constriction of blood vessels in the finger. The constricted vessels contain less blood, so the level of water falls in the plethysmograph.

What other conditions might change the blood flow and blood volume in the finger? Try them.

Chapter **34**

Veins and
venous pressure

Venous pressure is surprisingly low. When a person lies down, it is 5 to 10 mm Hg (millimeters of mercury) in most parts of the body and 0 mm Hg at the entrance to the heart. Contrast this with the typical arterial pressure of 120/75 mm Hg.

When a person stands, the venous pressure in the head and neck decreases, while that in the legs and feet increases. Gravity is the cause. By pulling blood down from the head, gravity lowers the venous pressure there to about −10 mm Hg, causing the veins of the scalp partially to collapse. In contrast, by pulling blood down from the heart, gravity raises venous pressure in the feet to as high as 90 mm Hg, causing veins of the feet to swell.

When a person stands, the gravity-induced pressure in the legs and feet forces fluids through capillary walls into the space outside the capillaries. Heat further increases the loss of fluids by causing the blood vessels to dilate, making them more permeable. For these reasons and because gravity pulls blood from the brain, people who stand for a long time in the heat may faint.

Fortunately, there are one-way valves in the veins of the arms and legs that allow blood to flow in one direction only—toward the heart. Movements of the arms or legs compress the veins, squeezing blood through the valves toward the heart. Then muscular relaxation allows blood to refill the veins, ready to be pushed through on the next contraction. In other words, the muscles of the legs and arms pump blood through valves of the veins somewhat as the heart pumps blood through the arteries. During movement, these muscles act as a "venous pump" (FIG. 34-1).

Some people have stand-up jobs that increase venous pressures in the legs. Also, pregnant women have high venous pressures in the legs because the fetus presses against the mother's veins, hindering movement of venous blood to the heart. When the venous pressures remain high for too long, some people

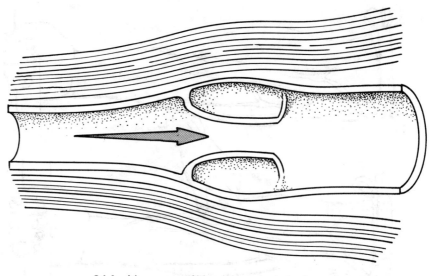

34-1   Movement of blood through a valve in a vein

develop permanently distended *varicose veins*. The distension separates the valves to such an extent that they no longer maintain an adequate flow of blood.

Materials

- Tourniquet
- Someone who has prominent veins in the arms

DIRECTION OF VENOUS BLOOD FLOW

In the early 1600s, William Harvey did an experiment to show the direction of venous blood flow. This experiment helped prove that blood circulates—that is, it moves from the heart to the arteries to the veins and back to the heart. Until then, blood was thought to shift back and forth in the vessels, somewhat as tides move at sea.

Today, you will repeat Harvey's experiment. Choose a subject who has well-marked veins in the arms. Tie a tourniquet snugly around this person's upper arm, but not so snugly that it is painful. A tourniquet is a band of cloth or rubber pulled tight enough to slow or stop bleeding. The pressure of the tourniquet will prevent venous blood from leaving the arm, causing the veins of the arm to swell. Notice that the tubular veins have occasional enlargements along their route. These enlargements are the sites of the one-way valves (FIG. 34-2).

Now press your finger on one of the veins, and as you continue pressing, move the finger toward the upper arm. Notice that blood flows into the vein behind your finger, refilling it. Thus the one-way valves of the vein must point in a direction that leads blood to the upper arm and eventually to the heart.

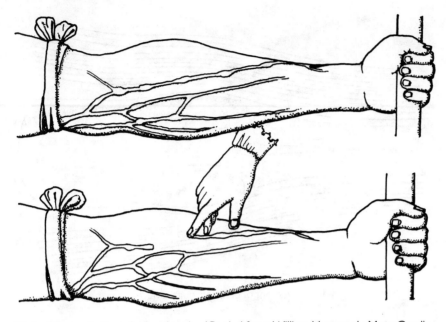

34-2   Demonstration of valves in veins (Copied from William Harvey, de Motu Cordis, 1628, opp. p. 56.)

Press again on the vein, but move your finger downward along the arm. Notice that the vein collapses and stays collapsed as you move toward the hand. The vein does not refill with blood (except at venous branches) because blood can move only toward the heart through the one-way valves. It cannot flow backward.

CAUTION: Experiment for two or three minutes only, then remove the tourniquet. Your subject will be eager to stop.

Materials

- Someone who has prominent veins in the feet

EFFECT OF EXERCISE ON VENOUS BLOOD FLOW

Choose a subject who has prominent veins in the feet. Your own feet may suffice. Have the subject remove shoes and socks and stand still for one minute. This will allow time for the veins to swell as gravity prevents blood from moving readily toward the heart. The venous pressure in the feet rises to about 90 millimeters of mercury (mm Hg).

Now ask the subject to take 8 or 10 quick steps in place and then again to stand still. *Immediately* look at his or her feet. Have the veins collapsed? Does stepping reduce venous pressure in the feet? Do the veins soon refill with blood? The answers are all yes. While stepping, the movements of the legs squeeze blood upward through one-way valves, reducing the venous pressure in the legs and feet. Consequently, the veins flatten and remain so for a brief time after exercise. But capillary blood continues to enter the veins, and as your subject is standing, gravity holds much of it in the feet. The veins, therefore, refill with blood.

Materials

- A meterstick or ruler
- Someone with prominent veins in the hands

VENOUS PRESSURE IN THE HAND

Choose a subject with prominent veins, perhaps you, and ask this subject to stand with arms hanging downward for one minute. This will allow time for veins in the hand to swell with blood. Now have the subject raise one arm slowly to eye level. Watch carefully. At what level do the veins of the hand collapse? Measure the approximate vertical distance from the heart to the place at which the veins collapse. In most people, this distance is about 10 centimeters, indicating that the venous pressure is 10 centimeters of blood. This pressure of blood equals 8 millimeters of mercury (mm Hg).

Materials

- Mirror

EFFECT OF CHEST COMPRESSION ON VENOUS BLOOD FLOW

Sometimes people hold their breath and strain, as in defecating and in women giving birth to children. The compression of the chest that accompanies this maneuver interferes with the return of venous blood to the chest and heart. It causes the venous blood to back up, raising its pressure.

To observe this effect, stand in front of a mirror, hold your breath, and strain—that is, attempt to exhale air forcefully while actually holding your breath.

CAUTION: Do this only if you are young and in good health, and continue for 15 seconds or less.

As you continue to strain, you will notice the veins enlarging in your neck. Arterial blood continues to enter these veins but cannot all escape to the heart.

While you are watching a concert someday, as on television, watch the veins in the necks of the performers. Those who play a series of long, high notes on trumpets or other wind instruments, will develop swollen veins. They are compressing their chests forcefully to deliver the notes, backing up venous blood in the process.

Also watch weight lifters as they lift the heaviest weights they can. In doing so, they subconsciously hold the breath to get a stable anchorage for their contracting muscles. The veins swell because compression of the chest interferes with venous blood flow.

OTHER ACTIVITIES

William Harvey's experiments on the heart and vessels provide a model that you might want to emulate. Because his studies were the first to show that blood circulates, his book on *The Motions of the Heart and Blood in Animals* has been repeatedly reprinted from the 1600s onward, and is often available in college and university libraries. By reading it, you can get ideas for further experiments.

Chapter **35**

Blood flow
through capillaries

T he capillaries were so crowded in fish tails, frog webs, and human fingers that Antoni van Leeuwenhoek, a seventeenth-century microscopist, concluded, "In a part of our skin, the size of a nail, as many as a thousand circulations . . . take place."[1] Later observers estimated that a human has ten billion capillaries.

Capillaries are many and small. Indeed, they are the smallest of all vessels, having tubes so narrow that blood cells file through one by one. Because the capillaries are many and because each red blood cell touches all sides of its capillary, nutrients and oxygen diffuse readily through the walls to nearby cells.

Blood enters the capillaries through contractile *arterioles* and leaves through *venules*. The arterioles dominate blood flow because arterioles have muscular walls. The capillaries, in contrast, have no muscle and the venules very little. When arteriole muscle contracts, the blood flow in capillaries slows or stops.

Blood flow to the tissues is cyclic, changing with need. This change results from constriction (narrowing) and dilation (widening) of blood vessels. Nerves control the constriction of arteries and large arterioles; oxygen and carbon dioxide control that of small arterioles and venules.

There are more capillaries than arterioles, but the ratio of the two depends on their location. Capillaries are most abundant in periodically active tissues, as in the muscles that move bones. August Krogh saw this and said, "The idea suddenly struck me that during rest, only a certain small fraction of the capillaries, suitably distributed, are open to the passage of blood, while all the rest are closed; and that increasing numbers are opened with work."[2] Krogh was right. Arterioles in muscles do dilate during exercise, so the capillaries open up—sometimes all of them. Thereby the muscles receive abundant nutrients and oxygen as they work.

Why do capillaries open? Krogh reasoned that a lack of oxygen might cause arteriole walls to relax, bringing more blood to the capillaries. We know from experiments that reduced oxygen causes blood vessels to dilate in the heart. We know further that if we shut off the blood supply of an arm, then suddenly let the blood back in, the arm flushes as dilated vessels fill with blood. Certainly oxygen decreases during hard work and in the occluded arm, but simultaneously carbon dioxide and other wastes increase. All of these changes promote dilation and an opening of capillaries.

Materials

- Goldfish
- Petri dish or other transparent dish or bowl
- Gauze or cotton
- Water from an aquarium or pond

BLOOD CIRCULATION IN A GOLDFISH

Pour several millimeters' depth of aquarium or pond water into a petri dish or similar transparent container. Put a small goldfish in the dish, and cover it—except the tail—with a layer of gauze or cotton soaked in the water (FIG. 35-1). Then put the petri dish on the stage of a microscope, and direct a light through the tail. If the fish flips its tail, move the dish and refocus.

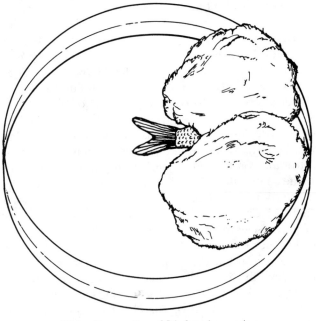

35-1   Preparation of fish for observation

Using low power, look down the microscope at an unforgettable display of streaming blood cells in a lacework of vessels. Then, if the fish is calm, try high power for a better view of the blood cells (FIG. 35-2). Look for 10 to 15 minutes, then return the fish to fresh water.

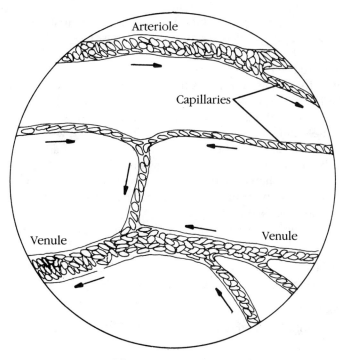

35-2   Microcirculation in the tail of a fish

Distinguish arterioles from capillaries from venules. If the heart beats strongly, you may tell large arterioles by the rhythmic spurts of blood gushing through them. But usually the pulse weakens and disappears by the time it reaches the arterioles of the tail, so the flow appears smooth. If the flow is smooth, identify arterioles by looking at the branches of smaller arterioles and capillaries coming off them. If the blood cells are moving from a larger vessel into smaller vessels, the larger vessel is an arteriole. Arterioles always branch into smaller and smaller arterioles and then into capillaries. If instead the blood cells are moving from smaller vessels into a larger vessel, the larger vessel is a venule. Venules always collect blood from smaller venules or capillaries. Capillaries are the narrow, thin-walled vessels located between the arterioles and the venules. They are so narrow that blood cells file through them one by one.

A distant heart drives blood cells to their capillary destiny. The pace usually slows as the cells pass through smaller and smaller arterioles into capillaries. Can you see the difference? The pace slows because there are many more capillaries than arterioles. The combination of all capillaries has a larger cross-sectional area than does the combination of all arterioles. The blood slows as it moves into the capillaries with their larger cross-sectional area, just as water slows in passing from a stream into a lake. Then the blood speeds as it moves from capillaries into venules because the combined cross-sectional area of capillaries is larger than that of venules.

Look closer at individual red blood cells. What are their shapes? Are they flexible in passing around sharp bends in the vessels? Judging by their speed through capillaries, does oxygen and carbon dioxide exchange occur quickly? How long is a red cell in its capillary?

Examine the venules further. Is the flow faster in their centers or along their walls? To see this, look at the speed of the moving cells in a venule wide enough to hold several cells side by side. Friction occurs where the cells touch the wall of the venule. The friction slows movement.

Look also at the skeletal arrangement in the fin and at the gold and sometimes black pigment cells. When the fish is light in color, the black cells are nearly spherical, showing little of their pigment. When the fish is dark, the pigment extends into branches, spreading the color. The spherical cells then resemble plants—black, branching plants.

When you finish, return the fish to its aquarium or pond.

Materials

- Faucet
- Y connector
- Rubber tubing to fit them

THE EFFECTS OF CONSTRICTION AND DILATION ON FLOW

Constriction and dilation shift blood from one part of the body to another. You may show this effect by slipping a short piece of rubber tube over a faucet. Insert a Y connector into the free end of the tube. Then put short rubber tubes on the two free ends of the connector.

Turn on the faucet. Note that the flow of water is about the same through the two openings. Now use your thumb and forefinger to partially constrict one of the tubes from the connector. What happens to the flow from both tubes and why? What would happen in the body if you similarly constricted one of two arterial branches? How would this affect the blood flow in capillaries supplied by these arteries?

OTHER ACTIVITIES

In the previous goldfish or another, try to find an arteriole that pulsates. Count and record the rate. This rate corresponds to that of the heartbeat. Also, measure and record the breathing rate—as shown by movements of the mouth and gills—and, if you have a thermometer, measure and record the temperature of the water.

Now place ice cubes on the cotton and in the water that surrounds the fish, and allow 10 minutes for the fish to cool. Again measure the temperature of the water. What effect does cold have on breathing and thus on the transfer of oxygen to blood? What effect does it have on heartbeats and blood circulation? Slow breathing and slow heartbeats suffice because a cold animal needs little oxygen.

Endnotes

1. Leeuwenhoek, Antoni van. Quoted by J.F. Fulton and L. G. Wilson. *Selected Readings in the History of Physiology*, 2nd ed., p. 73. Springfield, Illinois: Charles C. Thomas, 1966.

2. Krogh, August. *The Anatomy and Physiology of Capillaries*, 2nd ed. New Haven: Yale University Press, 1929.

Part 10

Lungs and breathing

Chapter 36

Lungs and breathing

*T*wo kinds of muscle control breathing at rest. One is the *diaphragm*, the dome-shaped sheet of muscle that separates the chest from the abdomen. When the diaphragm contracts, its dome moves downward, drawing air into the lungs. The other muscle is the *intercostal muscles* between the ribs. When the external layer of intercostal muscles contracts, it pulls the ribs upward and outward, drawing more air into the lungs. Without the muscles, the lungs are helpless.

As air enters, it passes through smaller and smaller tubes to reach the lungs. Starting from the nasal cavities, it goes through the *pharynx* to the *larynx*, the site of the *vocal cords*. The movement of these cords produces speech. Then it goes through the windpipe, or *trachea*, a tube lined with C-shaped cartilages that is continuous with the larynx. You can feel the larynx and trachea if you move your fingers along the front of your neck. The trachea branches into two, tubular *bronchi* which branch repeatedly like the limbs of a tree before continuing as tubular *bronchioles*. Air goes through smaller and smaller bronchi and bronchioles to the *alveoli*.

The alveoli are tiny, bubble-like air sacs, the microscopic destination of inhaled air. The exchange of gases occurs here. Blood pours through capillaries of the alveoli like a stream over stones. Oxygen diffuses inward from the alveoli to the blood, and carbon dioxide diffuses outward from the blood to the alveoli.

Normal exhalation is passive. The muscles for inhalation relax, causing the ribs and diaphragm to return to their resting positions. But during exercise, additional air must be driven from the lungs. The internal layer of intercostal muscles contracts, pulling the ribs downward and inward, forcing out air. Also, the abdominal muscles tense, pushing the diaphragm upward, forcing out air.

Materials

- Lungs and trachea (windpipe) of a mammal
- Rubber or plastic tube
- String
- Compressed air
- Hand lens

EXAMINING AND INFLATING LUNGS

This study requires the lungs of a rabbit, hog, or other mammal raised for meat (FIG. 36-1). For these, contact a custom butcher or meat packer listed in the yellow pages of your phone book. Get the lungs and trachea (windpipe) intact.

Upon receipt, wash as much blood as possible from the lungs. Do the lungs float in water? Once an animal takes its first breath, gas stays in the alveoli, never to be fully expelled. Only the lungs of unborn animals sink.

Feel the spongy tissue. Imagine the immensity of its surface. In human lungs, the area for gas exchange is about 100 square meters—approximately the size of a badminton court. Every pulse of blood from the heart to the lungs spreads over this entire area.

How many lobes are present in the lungs of your animal? Are the same number present on both sides? In humans, there are two lobes on the left, three on the right.

Feel the tough, C-shaped pieces of cartilage in the trachea. Follow them up to the larynx and down to the branching bronchi. Look deep into the open end of the larynx to locate the vocal cords. These are the narrow, cord-like shelves on each side of an opening about halfway down the larynx. The opening is the *glottis*.

Insert a rubber or plastic tube into the larynx. Rhythmically blow air through the tube into the lungs, causing them rhythmically to inflate. Stop blowing and they deflate. These movements resemble those of breathing.

You can permanently inflate and dry the animal's lungs if you have a source of compressed air. Attach one end of a somewhat flexible rubber or plastic tube to the air outlet, and insert the other end into the animal's larynx. Gently push the tube through the larynx into the trachea, and tie it in place. Turn on a slow, continuous flow of air. The lungs will inflate. Let air continue to flow through for 12 to 48 hours, depending on the size of the lungs, until they have thoroughly dried. Keep the lungs maximally inflated during this time.

When you finish, examine the lungs. Notice how light they are. Cut off a small piece of one lung to inspect its holey appearance. Use a hand lens to identify the sturdy tubular openings of the bronchioles and the tiny, thin-walled, bubble-like alveoli. In life, oxygen diffuses through the alveoli to the blood; carbon dioxide diffuses from the blood to the alveoli.

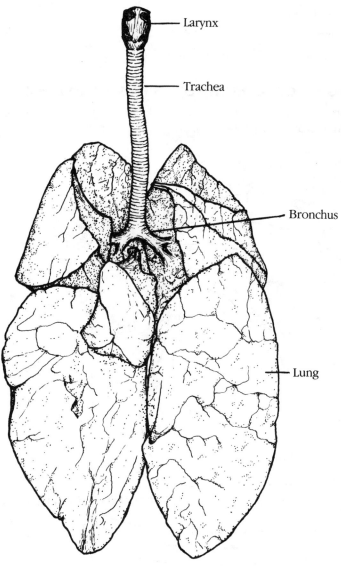

Larynx

Trachea

Bronchus

Lung

36-1 Lungs and trachea of a hog

Materials

- Bell jar or plastic jug
- Two balloons
- 1-hole rubber stopper
- Plastic Y-tube
- Small cork or bead

- Adhesive tape
- String
- Sheet of latex rubber to cover the bottom of the bell jar or jug

MAKING A MODEL OF THE DIAPHRAGM AND LUNGS

In this project, you will make a device that lets you imitate the movements of the diaphragm and lungs (FIG. 36-2).

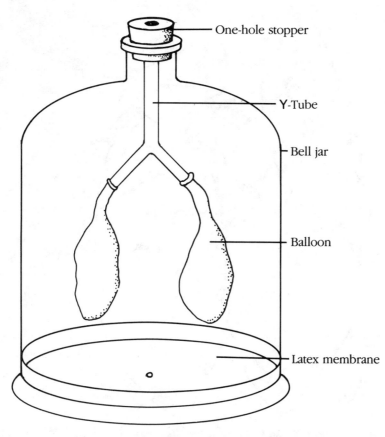

36-2   Model of the lungs and diaphragm

Plug a 1-hole rubber stopper into the opening at the top of a bell jar or that of a large, thick-walled, transparent or translucent plastic jug. Then attach a stretched, previously inflated balloon to each of the two prongs of a plastic Y-tube. Insert the free end of the Y-tube into the hole of the stopper. Cut a sheet of latex rubber to form a circular cover for the bottom of the bell jar. If you use a plastic jug, carefully cut off its bottom with a knife or saw, leaving a large, circular opening for the sheet of latex rubber. Press a small cork or bead into the middle of the sheet of latex, and tie it there to form a knob to pull on. Tape the latex to the bottom of the jar or jug. The latexmembrane represents the diaphragm; the balloons represent the lungs.

Pull on the knob. As you move the diaphragm downward, pressure decreases in the jar or jug, drawing air through the Y-tube into the balloons. This effect resembles inhalation. As the muscle of the mammalian diaphragm contracts, the diaphragm moves downward, drawing air through the trachea and bronchi to the lungs.

Release the knob. This allows the diaphragm to move upward, forcing air out of the balloons. This effect resembles exhalation. As the muscle of the mammalian diaphragm relaxes, the diaphragm moves upward, forcing air out of the lungs.

To illustrate active exhalation, push upward on the rubber diaphragm. The resulting increase of air pressure in the jar or jug forces additional air out of the balloons, causing the balloons to collapse. Similarly, mammals exhale deeply by contracting their abdominal muscles, pressing their abdominal contents and diaphragm upward. This movement forces additional air from their lungs but does not collapse them. However forcefully they exhale, there is always some air left in the lungs.

Materials

- Mirror

ACTIVATING THE MUSCLES OF BREATHING

You might want to stand bare-chested in front of a bedroom mirror while performing these exercises.

Inhale deeply. Notice that your rib cage is drawn upward and outward. You can see this in the mirror and feel it by placing your hands on your chest. The *external intercostal muscles* between your ribs contract at the same time the diaphragm contracts (FIG. 36-3). These intercostals draw the ribs upward and outward as the diaphragm moves downward, drawing air into the lungs.

If you inhale maximally, you can also see and feel the *sternocleidomastoid* muscles contracting in your neck. There are two of these muscles. They originate on the mastoid processes of the skull, bony swellings that you can feel immediately behind the earlobes. The muscles go from these bones to the upper end of the sternum (breastbone) where they insert. Put your hands on the two sternocleidomastoid muscles of your neck, below the mastoid processes. Now inhale as deeply as possible. Feel and see the two muscles contract toward the end of inhalation. By contracting, they pull the sternum and the attached ribs higher than would otherwise be possible, drawing in more air.

Exhale deeply. Watch your rib cage move downward and inward. *Internal intercostal muscles* pull the ribs downward, forcing air out of the lungs. Simultaneously, the diaphragm relaxes, moving upward to force additional air from the lungs.

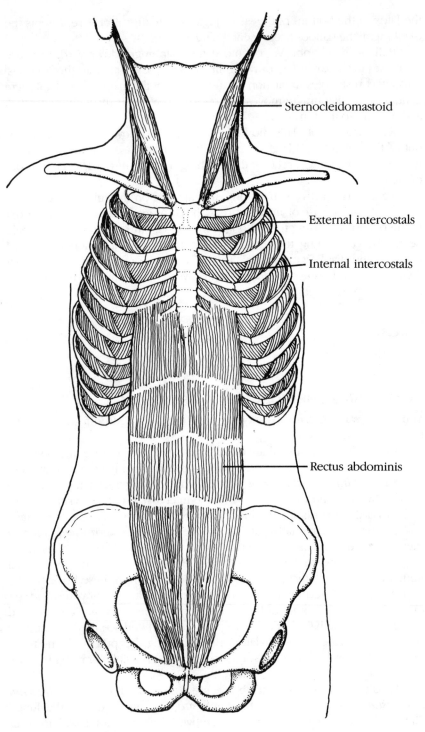

36-3   Muscles that contract to produce breathing

Exhaling maximally, you can also see and feel the abdominal muscles contract—especially the *rectus abdominis* that runs vertically along the midline. These contractions force the abdominal contents upward against the diaphragm, pressing it further upward against the lungs. The pressure moves additional air out of the lungs.

Materials

- Stethoscope
- Alcohol

LISTENING TO BREATHING

Wipe the earplugs of a stethoscope with alcohol, and insert them in your ears. Place the diaphragm of the stethoscope over your trachea (or that of another person). Listen to the air rush in and out of the trachea and bronchi as you breathe.

Now move the stethoscope to one or more of the intercostal spaces between your ribs. Listen again. The sound now is muffled, like the rustle of leaves by a gentle breeze, probably from air filling and leaving the alveolar sacs. If the bronchioles and alveoli are constricted or filled with extraneous matter—such as excessive mucus in pneumonia—the sound will instead be rasping, gurgling, or wheezing.

Materials

- Deep pan or bucket
- 4-liter (1-gallon) or larger bottle
- 1-liter flask or 1-quart bottle
- Plastic or paper plate
- Rubber hose with a diameter of about 2.5 centimeters (1 inch)

MEASURING AIR VOLUMES

Obtain a bottle that holds 4 liters (1 gallon) or more of fluid (FIG. 36-4). Fill it with 1 liter of water from a volumetric flask (or with 1 quart from a quart bottle). Make a mark or place a strip of adhesive tape at the 1-liter level. Continue filling the bottle with water, making a mark for each liter (or quart). A 1-gallon bottle holds approximately 4 liters. Invert the filled bottle into a deep pan or bucket of water, supporting it to prevent tipping. Keep air bubbles out by covering the mouth of the bottle with a plastic or paper plate before inverting it.

Now get a 1-meter (3-foot) length of wide-diameter rubber tubing or garden hose, and you are ready for the test. Inhale a maximally deep breath of air. Then exhale all of this air through the tube into the bottle. Determine the

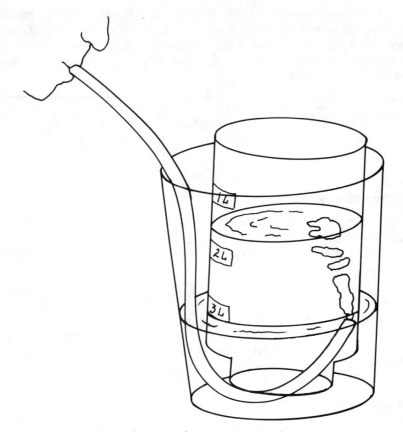

36-4   Apparatus for measuring air volumes

volume exhaled by checking the marks on the bottle. This is your *vital capacity*. Windy people will completely empty a 4-liter bottle.

In adults, the approximate vital capacity for women is 3.5 liters and that for men, 4.5 liters, but it varies considerably. In certain pulmonary diseases the capacity is less. For example, when the lungs are partly filled with fluid or when the muscles of respiration are paralyzed, there is less air or less ability to exhale it. The vital capacity, therefore, is reduced.

No matter how big-winded you are, you will never force all the air from the lungs. The lungs adhere closely to the chest wall, and the ribs prevent the chest from completely collapsing. The volume of air that remains in the lungs after maximal exhalation is called the *residual volume*. Normally it is about 1.5 liters. The residual volume contains enough oxygen to keep the blood oxygenated between breaths.

Fill the bottle again with water. Exhale a normal breath into it. How much water does this displace? In most adults, this *resting tidal volume*, as it is called, is about 0.5 liter (500 milliliters). Do not be alarmed if your tidal volume differs greatly from the average. You compensate for larger or smaller volumes by having a slower or faster breathing rate.

OTHER ACTIVITIES

Compare the differences in breathing rates of a person doing different exercises, such as walking, weight lifting, and bicycling at different rates. Also, compare the differences in vital capacities of men and women, and of persons who are more or less physically active.

Chapter **37**

Breath holding

Why does a swimmer dart to the surface after a minute or so underwater? And why does the swimmer breathe deeply upon surfacing? If you answer *excess of carbon dioxide* or *lack of oxygen*, you are correct. Carbon dioxide increases and oxygen decreases as the cells of the body consume nutrients. The increased carbon dioxide compels breathing through its effect on the *respiratory centers* of the *medulla*, or lower brain (FIG. 37-1). And the decreased oxygen activates chemoreceptors in arteries—specifically in the carotid arteries and aorta—receptors that in turn activate the respiratory centers.

Carbon dioxide and lack of oxygen are just two of several factors that regulate breath holding. The others are: conscious commands; motor signals from the cerebrum; receptor signals from contracting muscles; emotion; and acidity.

We know from experience that we can *consciously* stop breathing. To do this, we mentally signal the respiratory centers in the medulla to stop working. In contrast, by the other means listed, we signal the respiratory centers to start working. The start-working signals limit the time we can hold our breaths. For example, as we swim, the *motor* (movement-control) *areas* of our cerebrums stimulate the medulla at the same time they stimulate the muscles. Likewise, the *stretch receptors* in our contracting muscles—the sense organs that detect stretch —stimulate the medulla. Also, the *emotion* or excitement of swimming and *increased acidity* of blood stimulate the medulla. With all these stimuli acting on the respiratory centers of the medulla, we finally breathe.

The same stimuli control breathing in exercise. In water or on land, our cells use more oxygen and produce more carbon dioxide during exercise. We might expect, therefore, a big change in the carbon dioxide and oxygen content of the body, whatever the exercise. But this is not so. In most exercise, we do not hold our breath. Instead, we breathe rapidly and deeply. Thus we

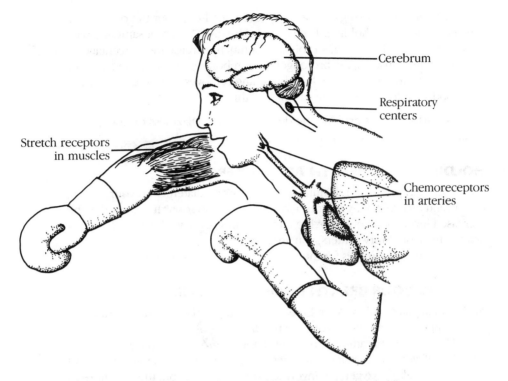

37-1 Parts that control breathing and breath holding

inhale a larger volume of oxygen than usual and exhale a larger volume of carbon dioxide. The rapid, deep breathing prevents any major decrease of oxygen or increase of carbon dioxide in the blood. The stimuli from these gases to the respiratory centers, therefore, are weaker than might be expected. By combining these stimuli, however, with others from the motor areas, stretch receptors, emotion, and acidity, we get tremendous increases in breathing.

Materials

- Watch or clock with a second hand

HOLDING YOUR BREATH

Begin these experiments by holding your breath, timing it, and recording it. Most people hold their breath ½ to 1 minute. By then, carbon dioxide strongly stimulates their respiratory centers, causing deep breathing to get rid of it. Allow 1 or 2 minutes for complete recovery.

HOLDING YOUR BREATH AFTER DEEP BREATHING

Rapidly take 8 to 10 very deep breaths—no more—then hold your breath, time it, and record it. Allow 1 or 2 minutes for recovery. Does the deep breathing lengthen your breath holding? Why?

Deep breathing brings more oxygen to your lungs, but the oxygen has little effect on breath holding. The blood is already 97 percent saturated with oxygen in normal breathing, so you cannot much increase the percentage by deep breathing. Instead, the prolonged breath holding results from a greater exhalation of carbon dioxide. Deep breathing removes carbon dioxide from the blood, taking away the stimulus for breathing.

CAUTION: Do not prolong the deep breathing. Dizziness or cramps can result.

HOLDING YOUR BREATH AFTER EXERCISE

Run in place rapidly for 1 minute, then stop. Immediately hold your breath, time it, and record it. Why do you start rebreathing so soon? If you say that the increased levels of carbon dioxide and acid and decreased levels of oxygen activate the respiratory centers, you are correct.

Allow several minutes for the recovery of normal breathing.

HOLDING YOUR BREATH DURING EXERCISE

Again run in place rapidly. After 45 seconds, hold your breath and time it while continuing to run rapidly. Record the result.

Why is your breath-holding time shorter *during* exercise than after exercise? If you say that increased levels of carbon dioxide and acid and decreased levels of oxygen are involved, you are correct, but there is more to it than this. Because you continued to run while holding your breath, the motor areas of your cerebrum and the stretch receptors in your contracting muscles activated the respiratory centers in your medulla. With so many stimuli to your medulla, you were quickly forced to breathe.

Allow several minutes for the recovery of normal breathing.

HOLDING YOUR BREATH AFTER INHALING OR EXHALING

Hold your breath after taking a maximal *inhalation*. Time and record it. When your lungs are fully inflated, do you have an urge to inhale or exhale? Which do you do first at the end of breath holding? In this experiment, the stretch receptors of deeply inflated lungs act through the medulla to cause exhalation.

Now hold your breath after taking a maximal *exhalation*. Time and record it. When your lungs are fully deflated, do you have the urge to inhale *or* exhale? Which do you do at the end of breath holding? The stretch receptors of deeply deflated lungs act through the medulla to cause inhalation.

Your breath-holding time, I suspect, was much shorter after exhalation than after inhalation. You quickly absorbed oxygen from your deflated lungs. Then the lack of oxygen excited the chemoreceptors in your arteries and medulla. You responded by quickly inhaling.

In strenuous exercise, the lungs alternate between deep inflation and deep deflation. In either case, stretch receptors in the lungs send signals to the respiratory centers that keep the lungs from overstretching. After deep inhalation, the receptors direct exhalation. After deep exhalation, the receptors direct inhalation.

OTHER ACTIVITIES

Compare the breath-holding times of people who exercise often with those who exercise seldom. Include one or two long-distance swimmers, if possible.

Chapter **38**

Carbon dioxide and oxygen

*I*n the early days of submarines, crews breathed only the air trapped inside their vessels or that outside when they surfaced. Sometimes the absence of ventilation caused sickness and death. This occurred, for example, in the S-4 that sank off Provincetown, Massachusetts, in 1927. The crew tapped messages to rescue ships 50 meters above, but a storm delayed the rescue. When finally reached, the entire crew of 40 had suffocated.

As the dying crew rebreathed the stale air, they added carbon dioxide to it and subtracted oxygen. Sufficient change in either gas can be fatal. Normal air contains 0.03 percent carbon dioxide. As the percentage of this gas increases, the breathing rate and depth of humans and other mammals increases. At 5 to 6 percent carbon dioxide, the breathing is extremely deep and labored. Headache and nausea develop. At 10 percent, the maximum tolerated, the air inhalation increases from a normal 8 liters per minute to an abnormal 70. Higher concentrations depress breathing, leading to unconsciousness and death.

The carbon dioxide in exhaled and rebreathed air comes from the breakdown of carbohydrates, fats, and proteins—our sources of energy. Each of the three nutrients contains carbon (C), oxygen (O), and hydrogen (H). During breakdown, these elements combine with the oxygen of air to produce carbon dioxide (CO_2), water, and energy. For example, in the breakdown of the carbohydrate glucose,

$$C_6H_{12}O_6 + 6\ O_2 \longrightarrow 6\ CO_2 + 6\ H_2O + \text{energy}$$
$$\text{glucose} \quad \text{oxygen} \qquad \text{carbon} \quad \text{water}$$
$$\text{dioxide}$$

We breathe, therefore, to get oxygen for generating energy, and to dispose of carbon dioxide.

The breakdown of nutrients occurs in cells from which carbon dioxide is carried to the lungs. Normally, exhaled air contains 4 percent carbon dioxide. During exercise, the production of the gas increases—as more nutrients are consumed to produce energy—but only small amounts accumulate inside the body. Rapid, deep breathing gets rid of the rest.

The *medulla* of the brain controls breathing (FIG. 37-1). Carbon dioxide acts here. A little increase in carbon dioxide—as when we rebreathe air from a confined space—causes an increase of medullar discharge to the muscles of the diaphragm and chest. These muscles, in response, cause the lungs to draw in and expel air more rapidly.

Materials

- Beaker or glass holding 250 milliliters (1 cup)
- Wide-diameter soda straw or other tube
- Watch or clock with a second hand
- Limewater (If you prepare your own limewater, you will also need calcium hydroxide and a glass container holding 500 milliliters (2 cups) or more.)

DETECTING CARBON DIOXIDE IN YOUR BREATH

Being careful, put 2 grams (½ teaspoon) of calcium hydroxide (slaked lime) in 500 milliliters (2 cups) of water. Stir the solution, cover the container, and let it stand overnight. In the morning you will have a saturated solution of calcium hydroxide called "limewater." The excess calcium hydroxide settles to the bottom of your container.

CAUTION: Calcium hydroxide is moderately caustic. Use adult supervision and wear safety goggles when making and using the solution. If you get hydroxide on your skin or clothes, wash it off with a large volume of running water.

Pour 200 to 250 milliliters (1 cup) of the clear limewater into a beaker or transparent drinking glass. Exhale at your normal rate through a wide-diameter soda straw or other tube into the solution. *Inhale through your nose*, but exhale directly into the beaker or glass. The carbon dioxide in your breath joins the calcium hydroxide in water to form calcium carbonate. The latter precipitates, making the solution cloudy. Record the time for the solution to become cloudy.

Empty and then refill the container with the same volume of limewater. Rapidly and vigorously exercise for 3 minutes or longer. Stop and immediately start exhaling into the solution. Again record the time it takes to become cloudy. The time should be less because more carbon dioxide is produced by your cells and expelled through your lungs during exercise.

When finished, rinse the beaker or glass repeatedly with water to remove the limewater and calcium carbonate.

Materials

- Human subject
- Paper bag

EFFECT OF CARBON DIOXIDE ON BREATHING

Count the respiratory rate of a resting subject for 1 minute, and observe the depth of breathing. Preferably do this when the subject is unaware of your counting. The breathing might seem shallow because a person at rest exhales only 400 to 600 milliliters of air in each breath.

Now have the subject breathe directly into a *paper* (not plastic!) bag for a few minutes or until he or she feels discomfort. Toward the end of this period, measure the breathing rate and observe its depth. Do the frequency and depth increase as carbon dioxide accumulates? What differences does the subject feel in breathing?

Carbon dioxide from the breath collects in the bag. It stimulates *chemoreceptors* in the brain's medulla and in the carotid arteries and aorta (FIG. 37-1). The carotid arteries are in the neck, and the aorta in the chest. The chemoreceptors then excite the respiratory centers of the medulla, causing an increase in the breathing rate and depth.

As the carbon dioxide in the bag increases, oxygen decreases. The subject consumes the oxygen to support life. When the level of oxygen drops sufficiently, it acts through the chemoreceptors of the carotid arteries and aorta to stimulate breathing. The chemoreceptors alert the respiratory centers in the medulla, and these centers—already responding to carbon dioxide—direct a further increase in the breathing rate and depth.

Materials

- Narrow soda straw or similar tube

EFFECT OF CONSTRICTED AIRWAYS ON BREATHING

Observe your normal breathing depth, and count the rate. Now begin breathing through a narrow soda straw or similar tube. Do you find this difficult? Imagine how breathing feels for persons who have asthma or other constrictions of their airways. Perhaps you have experienced such difficulties when you had a cold or bronchitis. Continue breathing for a few minutes through the narrow tube. This will allow carbon dioxide to collect in your cells, blood, and brain. How do you compensate? Now remove the tube from your mouth. The excess carbon dioxide in your body continues to stimulate the respiratory centers. How does it affect your breathing rate and depth? How does this change affect the removal of carbon dioxide from your body?

Materials

- Candle on a support that floats in water
- Shallow pan or bowl of water

- Large beaker or wide-mouthed bottle
- Watch or clock with a second hand
- Rubber tube or other flexible tube
- Soda lime (optional)

REMOVAL OF OXYGEN FROM AIR WHILE BREATHING

Because the body consumes oxygen during the breakdown of nutrients in cells, there is less oxygen in venous blood leaving the cells than in arterial blood entering them. Also, since the oxygen of venous blood is exhaled through the lungs, there is less oxygen in exhaled air than in inhaled air.

To demonstrate this, float a lighted candle in a shallow pan or bowl of water. Place an inverted beaker or wide-mouthed bottle over the candle (FIG. 38-1). Measure how long it takes to burn out and how far the water rises to replace oxygen as it is consumed. Air out the beaker, reinvert it in the water, and exhale a large volume of air into it through a flexible tube. Now relight the candle and insert it under the beaker, admitting as little fresh air as possible. The candle flickers and goes out quickly. Measure how quickly and how far the water rises. We conclude that part of the oxygen has been consumed by the body, since air leaving the lungs will not support the flame.

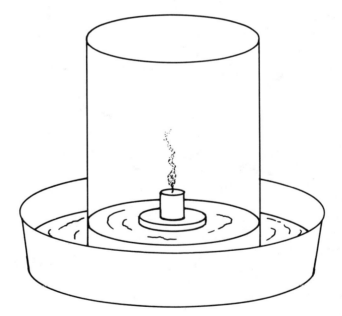

38-1   System for removing oxygen from air

This experiment does not account for the effect of carbon dioxide on a flame. If you want to eliminate this effect, float a container of soda lime in the inverted beaker during both tests. Soda lime absorbs carbon dioxide.

CAUTION: Because soda lime is alkaline, you should wear surgical gloves while handling it, and wash your hands afterward.

OTHER ACTIVITIES

Have a subject breathe through his or her mouth into a wide-diameter tube, such as a garden hose. Pinch the nose shut to allow breathing only through the mouth and tube. Carbon dioxide increases and oxygen decreases in the tube, stimulating respiration. Compare the effect of breathing through tubes of different lengths—for example, tubes of 3 centimeters' diameter that are 1, 2, and 4 meters long. *Use only one subject to prevent rebreathing of air exhaled from others.* The longer the tube, the more air the subject inhales to get fresh air. The increased inhalations are involuntary, regulated by the changes in carbon dioxide and oxygen.

Appendix **A**

The metric system

Scientists measure in tens, hundreds, and thousands with the international system of metric units. They avoid the awkward inches, ounces, and Fahrenheit degrees used by many people in the United States. The metric system has standard prefixes—such as *centi-* (hundredth) and *milli-* (thousandth)—preceding each measure. For example, one meter contains 100 *centimeters* and 1000 *millimeters*.

Some people juggle figures when converting one unit to another, depending heavily on calculators. Because there are 36 inches in a yard, for example, they multiply the number of yards by 36 to get inches. Scientists and other people simply move a decimal point. Because there are 100 centimeters in a meter, they multiply the number of meters by 100 to get centimeters; that is, they move the decimal point two digits to the right. They immediately see that a tube 2.5 meters long is also 250 centimeters long.

Similarly, some people multiply the number of quarts by 16 to get ounces. Scientists instead move a decimal point to convert liters to milliliters.

Also, some people use 32° and 212° Fahrenheit for the freezing and boiling points of water. Scientists instead use 0° and 100° Celsius for the same points.

See TABLE A-1 and the following list for useful conversions and reference points.

- A nickel (coin) weighs about 5 grams
- 2 level teaspoons of table salt (sodium chloride) weight about 9 grams
- A tablespoon holds about 15 milliliters or 0.5 fluid ounce
- There are 20 drops in 1 milliliter of water
- 1 milliliter of water weighs 1 gram

- Celsius degrees are ⅘ larger than Fahrenheit degrees
- Oral temperatures of humans are about 37° Celsius
- Room temperatures are about 20° to 25° Celsius

Table A-1   Measurements and Conversion Factors

Unit	Symbol	Value	Equivalent
Kilometer	km	1,000 m	0.62 mi
Meter	m	100 cm	39.37 in
		0.91 m	1 yd
Centimeter	cm	0.01 m	0.39 in
		2.54 cm	1 in
Millimeter	mm	0.001 m	0.039 in
Micrometer	μm	0.001 mm	1 micron
Kilogram	kg	1,000 g	2.2 lb
Gram	g	1,000 mg	0.035 oz
Milligram	mg	0.001 g	
Microgram	μg	0.001 mg	
Liter	L	1,000 ml	1.06 qt
		0.95 L	1 qt
Milliliter	ml	0.001 L	cm^3
Microliter	μL	0.001 ml	

To convert °C to °F: °F = 9/5 (°C) + 32
To convert °F to °C: °C = 5/9 (°F) − 32

Appendix **B**

Suppliers

*F*or the experiments, you may obtain most materials from your kitchen, backyard, or nearby fields and streams. Get others from supermarkets, meat markets, drugstores, hardware stores, aquarium shops, schools, and medical suppliers. For the latter, see "Medical Equipment and Supplies" in the yellow pages of your phone book.

If you cannot find materials locally, contact one of the companies listed below. All except Edmund have plants and animals, living and preserved. Also, most have chemicals and glassware. If you are a student, ask your teacher for suggestions and perhaps for the appropriate catalogs. Some companies only accept orders placed through schools.

Blue Spruce Biological Supply Co.
221 South St.
Castle Rock, CO 80104
(800) 621-8385

Carolina Biological Supply Co.
2700 York Rd.
Burlington, NC 27215
(800) 632-1231

Connecticut Valley Biological
82 Valley Rd., P.O. Box 326
Southhampton, MA 01073
(800) 628-7748

Edmund Scientific Supply Co.
101 E. Gloucester Pike
Barrington, NJ 08007-1380
(609) 573-6250

Fisher Scientific, EMD Division
4901 W. LeMoyne St.
Chicago, IL 60651
(800) 955-1177

Frey Scientific Co.
905 Hickory Ln., P.O. Box 8101
Mansfield, OH 44901-8101
(800) 225-FREY

Nasco
901 Janesville Ave., P.O. Box 901
Fort Atkinson, WI 53538-0901
(800) 558-9595
or
Nasco West
1524 Princeton Ave.
Modesto, CA 95352-3837
(800) 558-9595

Nebraska Scientific
3823 Leavenworth St.
Omaha, NE 68105
(800) 228-7117

Northwest Scientific Supply Co.
4311 Anthony Ct. # 700, P.O. Box 305
Rocklin, CA 95677
(916) 652-9674

Powell Laboratories Division
19355 McLoughlin Blvd.
Gladstone, OR 97027
(800) 547-1733

Sargent-Welch Scientific Co.
P.O. Box 1026
Skokie, IL 60076-8026
(800) 727-4368

Southern Biological Supply Co.
P.O. Box 368
McKenzie, TN 38201
(800) 748-8735

Ward's Natural Science Establishment
5100 West Henrietta Rd.,
P.O. Box 92912
Rochester, NY 14692-9978
(800) 962-2660

or

815 Fiero Ln., P.O. Box 5010
San Luis Obispo, CA 93403-5010
(800) 872-7289

Glossary

abduct To draw away from the midline or side of the body, as in raising an arm or leg sideways.

acetic acid (formula CH_3COOH) A colorless, acidic liquid that gives vinegar its sour taste. Vinegar is five percent acetic acid.

acetylcholine (abbreviated ACH) A chemical transmitter released from parasympathetic nerve cells, such as those entering the heart, blood vessels, intestine, and muscles.

active transport The movement of molecules and ions through cell membranes by means requiring metabolic energy, that is, the expenditure of calories.

adenosine triphosphate (abbreviated ATP) The main energy source for cells, such as those in exercising muscles. ATP is formed during the breakdown of glucose and other nutrients.

adrenaline A hormone secreted by the adrenal medulla and sympathetic nerves. Among other actions, it causes the heart to beat faster and blood vessels to constrict in the skin. Also called *epinephrine*.

agar gel A solid, jellylike substance upon which biologists grow microorganisms.

air bladder An internal, air-filled sac that helps a fish suspend itself in water, floating without effort. Lungfishes use it as a working lung. Also called *swim bladder*.

albino A plant or animal that lacks pigmentation, causing it to be white or colorless.

alga (plural algae) One of a group of mainly aquatic plants, such as seaweeds and pond scums.

alkaline Having the properties of an alkali. An alkali is a caustic substance that neutralizes acids.

allantois A pouch that grows from the fetal gut of reptiles, birds, and mammals. In reptiles and birds it serves as a respiratory organ and for the storage of wastes.

alveolus (plural alveoli)   A small cavity. In the lungs, one of the dome-shaped sacs through which gases diffuse between the air and the blood.

ammonium hydroxide (formula NH_4OH)   A corrosive, alkaline solution that is formed when ammonia dissolves in water.

amoeba (plural amoebas or amoebae)   A single-celled, transparent protozoan that crawls and feeds by extending lobes from its body.

amphibian   Any of a class of cold-blooded vertebrates that breathe with gills as larvae and with lungs as adults, such as frogs, toads, and salamanders.

amylase   An enzyme that digests carbohydrates.

antenna (plural antennae)   A movable sensory structure (feeler) that projects from the head of an insect or crustacean.

anterior   The head end of animals having four or more legs, and the front side of animals, such as humans, having two legs.

antidiuretic hormone (abbreviated ADH)   A hormone secreted by the hypothalamus and stored in the pituitary. ADH promotes water absorption from the kidneys, reducing the volume of urine excreted. Also called *vasopressin*.

anus   The opening through which intestinal wastes are excreted.

aorta   The large arterial trunk that carries oxygenated blood from the heart to arterial branches supplying the body.

apex   The narrow end or tip.

aphid   A small, oval, sluggish insect that sucks liquid from plants.

artery   Any of the thick-walled, branching vessels that carry blood from the heart to the arterioles.

arteriole   Any of the narrow but muscular vessels that carry blood from the arteries to the capillaries.

arteriosclerosis   A chronic disease that causes a hardening or stiffening of the arteries. It results from an accumulation of lipids, connective tissue, and calcium in arterial walls.

arthropod   An animal with jointed legs and a hard external covering. The arthropods include insects, spiders, crayfish, crabs, centipedes, and millipedes.

astigmatism   A blurring of vision resulting from an unequal curvature of the cornea or lens.

ATP   *See adenosine triphosphate.*

Atrioventricular valve (abbreviated AV valve)   A valve located between an atrium and a ventricle of the heart. The right AV valve is also called the *tricuspid* and the left the *mitral* or *bicuspid*.

atrium (plural atria)   A chamber providing an entrance to another structure. Usually refers to a chamber of the heart that passes blood from veins to a ventricle.

auditory canal   The tube leading from the external, flaplike pinna of the ear to the eardrum. The auditory canal transmits sound waves. Also called *external auditory meatus*.

auditory tube   A canal passing between the middle ear and the pharynx. When swallowing or yawning, the tube opens, allowing the air pressure of the middle ear to change to that of the outside. Also called *eustachian tube*.

AV valve   *See atrioventricular valve.*

bacterium (plural bacteria) A single-celled, microscopic organism which typically has a rigid cell wall but which lacks a membrane-bound nucleus. Some species cause disease.

balsam A resinous, oily liquid that flows from plants, such as the balsam fir.

basal metabolism The energy consumed and heat produced by an organism that is resting and fasting. It is usually determined by measuring the rate of oxygen consumption.

battery jar A glass container with straight sides and a round, square, or rectangular opening at the top.

beaker A deep, wide-mouthed, cuplike glass or plastic container for liquids.

beetle Any insect that has four wings, the outer two of which are hard and protective.

bell jar A glass or plastic jar that is shaped like a bell, open at the bottom and rounded at the top. It is designed to hold gases or a vacuum.

Betadine Trade name for povidone-iodine, an antiseptic available from pharmacists or medical suppliers.

bile A green fluid secreted by the liver, stored in the gallbladder, and released into the small intestine where it promotes the breakdown and absorption of fats.

brachial artery The major artery passing through the upper part of the arm.

bronchus (plural bronchi) Any of the branches of air tubes connecting the trachea to the bronchioles of the lungs.

bronchiole Any of the minute tubes that connect the bronchioles to the air sacs of lungs. In humans, bronchioles are one millimeter or less in diameter.

brownian movement An erratic movement of particles hit by molecules moving at rapid speeds.

buttocks The muscular seat of the body.

calorie The amount of heat required to raise one milliliter of water one degree Celsius. This unit is $\frac{1}{1000}$th of a kilocalorie.

calcium hydroxide (formula $Ca(OH)_2$) A white alkaline powder used to make limewater.

camera, single-lens-reflex A camera in which the photographer looks directly though the lens, seeing the exact view that will be taken on the film.

capillary Any of the thin-walled, extremely narrow blood vessels through which blood cells pass single file. Oxygen and nutrients diffuse from capillaries into surrounding tissues, and carbon dioxide diffuses from the tissues into the capillaries.

carbohydrate Any of the chemical compounds of carbon, hydrogen, and oxygen, such as sugars, starches, glycogen, and cellulose, usually having a 2:1 ratio of hydrogen to oxygen.

carbon dioxide (formula CO_2) A gas produced in the tissues as a by-product of metabolism and eliminated by the lungs.

carbon tetrachloride (abbreviated CCl_4) A colorless, nonflammable but toxic liquid that is used as a solvent and as an anesthetic for insects.

carbonic acid (formula H_2CO_3) A weak acid produced when carbon dioxide is dissolved in water.

cardiac muscle   The muscle that forms the wall of the heart.

cardiovascular   Pertaining to the heart and blood vessels.

cardiovascular center   The part of the medulla (of the brain) that regulates the heartbeat and flow of blood through vessels.

carotid artery   One of the arteries that has pressoreceptors and chemoreceptors that alter the blood circulation and breathing.

cartilage   A sturdy translucent skeleton found in sharks and other cartilaginous fishes and in parts of mammals, such as the larynx and trachea.

cataract   An opacity of the lens, causing partial or total blindness.

caterpillar   The wormlike larva of a moth or butterfly.

caustic   Corrosive and burning, causing destruction of living tissue.

cell membrane   The membrane covering the cell and separating it from other cells or fluids. Also called *plasma membrane.*

cell   A boxlike, usually microscopic unit of protoplasm. Many cells together form the tissues and organs of the body.

cellulose   A carbohydrate molecule formed by the union of thousands of glucose molecules in plants. Cellulose is the chief constituent of the cell walls in plants.

Celsius   A temperature scale in which water freezes at 0 degrees and boils at 100 degrees. One degree Celsius equals ⅘ degrees Fahrenheit. *See Appendix A.*

centipede   A long, somewhat flattened arthropod with many segments in its body, each segment having one pair of legs. The frontmost legs are modified as poisonous fangs.

centimeter   0.01 meter or 0.39 inches. *See Appendix A.*

cerebellum   The large, hindmost projection of the brain that helps regulate bodily movements and balance.

cerebral cortex   The surface layer of gray matter on the cerebrum of the brain. The cortex detects sensations, integrates thought, and directs movement.

cerebral hemisphere   One of the two halves (right and left) of the cerebrum of the brain.

cerebrum   The large, uppermost, convoluted part of the brain.

cervical   Relating to the neck.

chemoreceptors   A sensory nerve ending, cell, or organ that responds to specific chemical stimuli. For example, the chemoreceptors in the medulla of the brain respond to oxygen deficiencies. The medulla then directs deeper, faster breathing, helping to compensate for the lack of oxygen.

cicada   An insect with a broad heavy body and big lacy wings.

cilium (plural cilia)   Any of the hairlike extensions from the surface of certain cells. Oarlike movements of cilia propel cells, small organisms, or other matter forward.

ciliary muscle   A muscle, encircling the eyeball, whose contraction regulates the shape of the lens and hence the focus of light.

ciliary processes   Lobes of the ciliary body that project inward from the ciliary muscle toward the lens of the eye. Ciliary processes secrete aqueous humor and provide sites of attachment for threadlike fibers to the lens.

cloaca   The common passage into which the reproductive, urinary, and intestinal tracts empty in birds, reptiles, amphibians, and many fishes.

coccyx   The tailbone of fused vertebrae at the end of the spine.

cochlea   The snail-shaped organ of the inner ear that holds receptors for sound.

cochlear nerve   The nerve connecting the cochlea to the brain. The cochlear nerve transmits impulses that determine what is heard.

cocoon   The silken envelope in which the inactive pupa of an insect lies.

coelenterate   A phylum of invertebrate (boneless) animals that includes the jellyfishes, sea anemones, corals, and hydras.

collagen   The main protein in the white fibers of connective tissue, as in ligaments, tendons, and bones.

compost   Decayed organic matter that is used for fertilizer.

compound eye   An eye that is composed of hundreds to thousands of tiny, simpler eyes crowded together. Compound eyes are found in many arthropods, including insects.

concave   Having a rounded, depressed surface.

conditioned response   A response acquired through repetitious training. Also called *conditioned reflex*.

cone   Any of the cone-shaped receptor cells in the retina that are sensitive to bright colored light.

constrict   To become narrow, as when a pupil or a blood vessel gets smaller by constriction.

control   A comparison that makes an experiment valid. In testing a new drug, for example, experimenters give a placebo (such as a sugar pill) to some subjects and the real drug to others. Then they compare the results.

converge   To direct inward to a point, as when someone focuses sunlight beneath a hand lens.

convex   Having a surface that is rounded outward, as in the convex lens of an eye.

convolution   A folded or tortuous shape, as in a ridge on the surface of the brain.

cornea   The outer, transparent, rounded part of the eyeball that works with the lens to focus light on the retina.

coronary artery   One of the arteries that supplies oxygenated blood to the muscle of the heart.

coronary occlusion   A partial or total blockage of blood flow through a coronary artery, usually caused by lipid deposits or a blood clot.

cover glass   A thin piece of glass that is placed over a specimen before observing it under a microscope. Also called *cover slip*.

cranial nerve   Any of the nerves that attach directly to the brain.

crayfish   Any of the 10-legged, freshwater crustaceans that resemble lobsters. Also called *crawfish* or *crawdad*.

cross section   A piece of something cut at a right angle to its axis. A cross section through the trunk of a tree, for example, shows tree rings.

culture medium   A substance on which bacteria, fruit flies, or other small organisms grow.

cytoplasm   The part of the cell between the nucleus and the cell membrane.

damselfly   An insect related to the dragonfly. Unlike the dragonfly, the damselfly folds its wings over its back when resting.

Daphnia   A tiny, freshwater crustacean having a transparent body and long antennae. Also called *water flea*.

defecate   To discharge feces (intestinal wastes) through the anus.

deionized water   Water from which the minerals are removed.

diabetes   A disease (diabetes mellitus) in which there is a deficiency of insulin or a lack of responsiveness to it. Consequently, glucose collects in the blood and is excreted in the urine.

dialysis membrane   A semipermeable membrane that allows small molecules but not large molecules to pass through it. Dialysis tubing is made from a dialysis membrane.

diaphragm   1. The dome-shaped sheet of muscle between the thorax and abdomen that contracts to produce inhalation. 2. A metal sheet having one or more openings, located under the stage of a microscope; light is passed through the diaphragm to the specimen.

diastolic pressure   The lowest of the two blood pressures measured in large arteries (the other is systolic pressure). Diastolic pressure averages about 75 millimeters in young, resting humans. It occurs during diastole, the period of ventricular relaxation when no blood passes from the heart to the arteries.

differentiate   Become different from the original type, as when an embryonic structure develops a new structure and function.

diffuse   to spread out by diffusion.

diffusion   A heat-induced, random movement of molecules and ions that causes them to move away from the source.

dilate   To expand or widen, as when the pupil of an eye dilates in dim light or when an artery dilates to pass more blood.

distilled water   Water that has been heated, vaporized, and condensed to remove dissolved and suspended particles.

diverge   To spread or move outward, as in diverging light rays.

dominant   In genetics, a trait that occurs in the offspring even when it is present in only one parent.

dorsal   On or toward the back side of an animal.

dowel   A cylindrical, wooden rod.

droppings   The dung or excrement of an animal.

duct   A narrow tube, especially one that carries the secretions from a gland.

dysentery   An intestinal disorder causing diarrhea, usually caused by an infection.

electrode   A plate, needle, or pin used to electrically stimulate part of the body or to receive an electrical signal from the body.

electron microscope   A device in which a beam of electrons is focused on cells or other minute objects. The objects are greatly enlarged for display on a fluorescent screen or photographic plate.

embryo   An animal at an early stage of development, before it becomes a fetus. In humans, the embryonic stage lasts from implantation in the uterus to the tenth week.

endocrine gland A gland, such as the pituitary or thyroid, that secretes hormones into the blood or other fluid surrounding it.

enzyme One of the many specific proteins that temporarily combines with a biochemical to accelerate a chemical reaction.

epiglottis A protruding, flexible flap overhanging the larynx. While swallowing, the epiglottis bends down, helping to keep food and liquids out of the larynx.

esophagus The muscular tube that carries food and drinks from the pharynx to the stomach.

estrogen Any of the hormones that promote estrus (heat) in mammals and the development of female characteristics.

excrete To discharge feces, urine, or other wastes from the body.

excretion The process of excreting wastes.

fat-soluble Anything that dissolves in fat.

feces Waste (droppings or dung) discharged from the intestine.

feedback The return of corrective information to a control center. For example, when a human generates excessive heat, some of this heat is circulated back (fed back) to the brain which then directs the body to sweat.

fertile Capable of developing or of having offspring.

fetal pig An unborn pig with organs and other features similar to those of an adult.

fetus An unborn vertebrate animal that already has the physical appearance of its species. A human is called a fetus after about 10 weeks growth in the uterus.

filter paper A paper that is porous enough to let liquids and small particles but not large particles pass through.

flatworm A flat, soft-bodied worm.

forceps Surgical tweezers used to grasp and hold objects.

fovea The small, depressed part of the retina in which light rays are focused when a person looks directly at an object.

frigate bird Any of several species of long-winged seabirds.

gallbladder A small, usually green sac in which bile from the liver is stored until its release into the intestine.

gastric juice The liquid secreted by the stomach into the cavity of the stomach.

gastric gland Any of the small, tubular glands in the stomach wall. The gastric glands secrete hydrochloric acid and pepsin, an enzyme that digests protein.

gene A genetic unit found in the nucleus of a cell. Genes are passed from parents to offspring, helping to determine the characteristics of the offspring.

gizzard A muscular pouch in the digestive tract of a bird, used to grind food.

glottis The opening between the vocal cords.

glucose (formula $C_6H_{12}O_6$) A sugar from which the body gets much of its energy.

glycogen   A carbohydrate formed abundantly in the liver and muscles by the chemical union of glucose molecules. The breakdown of glycogen produces glucose and energy to operate the body.

goiter   An enlarged thyroid gland.

gram   One-thousand milligrams or 0.035 ounces. *See Appendix A.*

gut   A digestive cavity, such as that into which hydra takes food.

hardware cloth   A flexible screen of steel wire woven in a close mesh.

heat   In female mammals, a period in which they are sexually receptive to males. Also called *estrus*.

hemoglobin   The red, iron-containing component of red blood cells. The iron combines with oxygen, allowing oxygen to be carried in blood.

heron   A wading bird with a long neck, long bill, and long legs.

histology   Study of the microscopic structure of tissues.

homeostasis   Tendency of conditions in the body to stay almost constant.

honeydew   A sweet, watery solution secreted by aphids, mealybugs, and certain plants.

hormone   Any of the chemical messengers that control bodily functions. Hormones are produced by endocrine glands and usually carried by blood to the cells upon which they act.

humerus   The long bone of the upper arm.

hydra   A tiny, aquatic animal with a narrow trunk from which tentacles project at one end.

hydrochloric acid (formula HCl)   A strong acid secreted by glands in the stomach.

hydrogen peroxide (formula H_2O_2)   A colorless, foaming, antiseptic bleach.

hydroxyapatite (formula $Ca_{10}(PO_4)_6(OH)_2$)   The mineral in which calcium phosphate is stored in bones and teeth, making them hard.

hypothalamus   The part of the brain between the thalamus and pituitary; it helps regulate emotions, feeding, drinking, urination, body temperature, and hormone secretion.

hypothyroid   Having a low metabolism and sluggish behavior. The condition results from a deficiency of thyroid hormones.

immunity   Resistance to a disease, especially to an infective disease.

incision   1. The act of cutting into the body. 2. The wound made by cutting the body.

incubate   To keep an embryo in a proper, warm environment that will allow it to develop.

index finger   The finger next to the thumb.

inflammation   A tissue reaction to injury. It swells and becomes redder and hotter from the dilation of small vessels.

inner ear   A part of the ear located deep in the skull. The inner ear contains sense organs for hearing and balance.

insertion   The place where a muscle attaches to a movable bone or bones.

intestine   The long tube winding from the stomach to the anus. Food is digested and absorbed in the intestine. *See also small intestine and large intestine.*

invertebrates   Animals that lack backbones, for example, worms, insects, and starfishes.

iodine   A nonmetallic element needed for the formation of thyroid hormones.

ion   An atom or group of atoms that carry an electric charge. Sodium (Na$^+$) and chloride (Cl$^-$), for example, are ions.

isopropyl alcohol (formula C$_3$H$_8$O)   An alcohol commonly used as an antiseptic. Also called *rubbing alcohol*.

kangaroo rat   A nocturnal, burrowing desert rat that hops on its hind legs.

keel   A bony, midline extension from the sternum of a bird. It resembles the keel of a boat.

kidney tubule   Any of the microscopically narrow but long tubes that receive fluid filtering out of capillaries in the kidneys. The tubules carry this fluid forward, transforming it into urine.

kidney   The organ in which urine forms. Each of the two human kidneys are reddish-brown, bean-shaped, and located on the back side of the abdominal cavity.

kilocalorie (abbreviated kcal)   The heat required to warm one liter of water one degree Celsius. When referring to the energy in food, a kilocalorie is called a Calorie.

lactic acid   An acid produced by the metabolic breakdown of glucose in the absence of oxygen.

ladybug   Any of the small, colorful, almost hemispherical beetles that feed mainly on small insects and insect eggs.

large intestine   The part of the intestine containing the cecum, appendix, colon, and rectum.

larva (plural larvae)   Any of the immature, wingless, wormlike creatures that hatch from the eggs of insects and begin feeding. The larva of a housefly is a maggot and the larva of a moth is a caterpillar.

larynx   The cartilaginous enlargement of the respiratory tract in which the vocal cords are found. The larynx can be seen externally in men as a knob at the front of the neck.

lateral   Toward the right or left side of the body.

leafhopper   Any of the small, leaping insects that suck the juices from plants.

life cycle   The changes in form and activity of an organism during its lifetime.

ligament   A strong band of connective tissue that binds one bone to another.

ligate   To bind or tie.

limewater   *See calcium hydroxide.*

linear   In a straight line.

lipase   Any of the enzymes that digest fats.

lipid   Any of the fats or fatlike substances.

liter   One-thousand milliliters or 1.06 quarts. *See Appendix A.*

litmus   A dye or dyed paper that turns red in the presence of acids and blue in the presence of bases (alkali).

liver   A large, reddish-brown, abdominal organ that makes bile, stores glucose, detoxifies harmful substances, and is the site of many metabolic reactions.

liverwort   A plant resembling and related to mosses.

lizard   A reptile distinguished from snakes by movable eyelids and usually by two pairs of well-developed legs.

locution   A form of speech.

longitudinal   Running lengthwise.

Lugol's solution   A solution used to test for starch in food or leaves. To make it, dissolve 10 grams of potassium iodide in 100 milliliters of distilled water; then add 5 grams of iodine.

lungfish   Any of the fishes that have an air bladder developed as a working lung.

lymph nodes   Round or oval bodies, 1 to 25 millimeters long in humans, found along the course of lymphatic vessels. The nodes contain amoeba-like cells that engulf and destroy microorganisms.

lymphatic vessels   Any of the thin-walled capillaries and vessels that carry lymphatic fluid from the tissues into the subclavian veins of the neck.

lymphocyte   A kind of white blood cell that helps destroy infective microorganisms. Lymphocytes are found both in lymphatic and blood vessels.

lymphoid organs   The lymph nodes, tonsils, thymus, liver, spleen, and other organs that produce lymphocytes.

malaria   An infectious disease resulting from protozoan parasites transmitted by tropical mosquitoes. The victim has periodic chills and fevers.

mammary glands   The milk-producing glands in the breasts of female mammals.

marrow   The red or yellow soft tissue in the cavities of bones. The red marrow manufactures blood cells. The yellow marrow contains fat and connective tissue.

mayfly   A mainly aquatic insect. The delicate, lacy-winged adult emerges from the water to fly, mate, and die.

mealworm   The larva of various beetles that feeds on grain and its products. Mealworms are raised for fish bait and as feed for some animals.

mealybug   Any of the scale insects that have a white, powdery covering and that harm the plants on which they feed.

medial   Toward the midline of the body.

medulla (plural medullae or medullas)   1. The medulla oblongata at the posterior end of the brain adjoining the spinal cord. The medulla helps regulate breathing, blood circulation, digestion, speech, coughing, sneezing, chewing, swallowing, and vomiting. 2. The inner part of some organs, such as the adrenal gland.

melanin   A black pigment found in the skin, hair, eyes, and other parts of the body.

membrane   A thin, flexible sheet, such as the membrane that envelops a cell.

menses   The monthly period when the lining of the uterus disintegrates, releasing blood and cellular debris through the vagina. Also called *menstruation*.

menstrual cycle   The monthly changes in female hormones and reproductive organs of women. Each cycle ends with the menses.

mercury   A heavy, silver-white, poisonous liquid used in instruments that measure temperature and pressure.

metabolism   The processes in the body in which substances are assembled or broken down and energy consumed or released.

metabolic   Pertaining to metabolism.

metamorphosis   The developmental change from one stage to another, as when pupae become moths or tadpoles become frogs.

meter   One hundred centimeters or 39.37 inches. *See Appendix A.*

meterstick   A measuring stick that is one meter long and usually marked off in centimeters and millimeters.

metronome   A device that regularly marks time by loud ticking.

microbe   A microscopically small organism, such as a bacterium, especially one that causes disease. Also called *microorganism.*

middle ear   The air-filled, middle cavity of the ear that contains three tiny bones, the ossicles.

midline   A median or middle line that runs down the longitudinal axis of the body or some part of the body.

milligram   One-thousandth of a gram. *See Appendix A.*

milliliter   One-thousandth of a liter or 20 drops. Also called *cubic centimeter. See Appendix A.*

millipede   A long, cylindrical arthropod with many segments in its body, each segment having two pairs of legs.

mitochondrion (plural mitochondria)   One of the subcellular structures that makes ATP, providing a source of energy for the body.

mitral valve   *See atrioventricular valve.*

molecule   The smallest part of a substance that retains all its properties. Each molecule is composed of one or more atoms.

molt   To shed the outer covering (exoskeleton), as occurs periodically in growing insects and other arthropods.

motor area   The part of the cerebral cortex that controls movement. Also called *motor cortex.*

motor nerve   A nerve that controls muscular contraction or glandular secretion.

motor neuron   A nerve cell that controls muscular contraction or glandular secretion.

mounting board   A board upon which to pin dead insects.

mount   An object displayed on a slide for microscopic examination.

mucus   A viscid, slimy secretion, such as that in the nose.

muscle fiber   Any of the long, threadlike muscle cells found in skeletal and heart muscle.

muslin   A plain-woven, bleached or unbleached sheet of cotton fabric.

mutant   An individual produced by mutation.

mutation   A rare change in genetic structure that makes the offspring different from the parents. Once produced, the mutation can be inherited by future generations.

myocardium   The muscle of the heart.

neuromuscular system   The brain and nerves and the muscles they cause to contract.

nucleus (plural nuclei)   The large membrane-enclosed structure near the center of the cell. It contains genes that control reproduction.

nymph   The larva of certain insects, such as dragonflies and mayflies.

objective lens   In a microscope or telescope, the lens that is closest to the object being viewed.

olfactory   Concerned with the sense of smell.

opaque   Impervious to light. An opaque lens, for example, appears cloudy.

opacity   The state of being opaque.

ophthalmologist   A medical doctor who treats diseases of the eye.

optic nerve   The nerve through which visual information passes from an eye to the brain.

order   A group of related plants or animals. For example, all moths and butterflies belong to the order Lepidoptera.

organism   A living creature.

origin   The place where a muscle attaches to an immovable or slightly movable bone or bones.

oscillate   To vary above or below a central level.

osmosis   The diffusion of water (or other solvent) through a semipermeable membrane from the side where water (or solvent) is more concentrated to the side where it is less concentrated.

ovary   A female reproductive organ in which eggs and female hormones (estrogens and progesterone) are produced.

ovum (plural ova)   An egg; the female reproductive cell.

oxygen (symbol O)   An element that is a colorless, odorless gas in air and that joins other elements in chemical compounds, such as water (H_2O) and glucose ($C_6H_{12}O_6$).

pacemaker   Something that sets the pace. For example, the sinoatrial node, serving as a pacemaker, directs the human heart to beat about 75 times per minute.

pancreas   A large gland, behind the stomach, that secretes digestive juices and two hormones, insulin and glucagon.

parasite   A sometimes harmful organism, such as a flea or tapeworm, that lives on or in another organism.

parasympathetic nerve   One of the nerves that originates in the brain or sacral region of the spinal cord and goes to various organs and glands. Activation of parasympathetic nerves causes the heartbeat to slow, respiratory passageways to constrict, and digestive glands to secrete.

patellar ligament   The strong band of connective tissue binding the patella (knee bone) to the tibia (shin bone).

pelvic bone   Hip bone.

pelvis   The bony structure formed by the pelvic bones (hip bones) and sacrum (the fused bone at the back of the hip), or the cavity formed by these bones.

penis   The bodily extension through which urine and sperm pass from the male.

peripheral   At the periphery, that is, away from the center.

petri dish   A shallow, circular, glass or plastic container and its cover, commonly used to hold media on which bacteria are grown.

petrolatum   An odorless, tasteless jelly that is used in ointments. Also called *Vaseline* or *petroleum jelly*.

phagocyte An amoebalike cell, such as a white blood cell, that engulfs and feeds upon microbes.

pharynx A muscular passageway in the throat that channels air into the larynx and food into the esophagus.

phenolphthalein An indicator that becomes red in the presence of bases.

photoreceptor Any of the receptor cells for light, as in the retina of an eye.

pipette or pipet A long, narrow, glass or plastic tube that is graduated to show the volume of fluid contained.

pituitary A small endocrine gland connected to the underside of the brain. It secretes hormones that affect growth, metabolism, reproduction, pigmentation, and urine formation.

plasma The liquid of blood, the carrier of blood cells.

plaster of Paris A white powder. When water is mixed with plaster, the resulting paste solidifies, allowing experimenters to make casts of animal tracks.

plethysmograph A device used to measure changes in the volume of an organ or limb, for example, changes in the blood volume of a finger induced by constriction and dilation of its blood vessels.

pneumonia Inflammation of the lungs caused by bacteria, viruses, or other irritants.

pocket mouse Any of the small, nocturnal, burrowing rodents that resemble mice but have longer hind legs. They mainly live in deserts.

pollen The dustlike spores (reproductive bodies) of seed plants.

posterior The tail end of animals having four or more legs, and the back side of animals, such as humans, having two legs.

predator An animal that hunts, kills, and eats other animals.

pressoreceptor Any of the stretch-sensitive nerve endings in the walls of the aorta and carotid arteries that detect changes in blood pressure. Acting through the brain and autonomic nerves, they help stabilize this pressure. Also called *baroreceptor*.

proprioception The reception of information with proprioceptors.

proprioceptor Any of the stretch-sensitive sensory structures in muscles, tendons, joints, and the inner ear that give information about body positions and movements.

proteinase Any of the enzymes that digest proteins.

protein Any of the chemical unions of a large number of amino acids in a single molecule.

protoplasm The organized complex of chemicals, nucleus, mitochondria, and other components that make up living cells.

protozoan (plural protozoa) Any of a subkingdom or phylum of one-celled, microscopic organisms, such as amoeba and paramecium.

pseudopod A temporary, lobular extension of protoplasm used by amoebas and certain white blood cells in moving and feeding.

ptarmigan One of the grouses of northern climates that have completely feathered feet.

pupa (plural pupae or pupas) An intermediate, postlarval stage in the development of some insects, such as moths, fruit flies, and ants. The physically inactive pupa later becomes an active adult.

receptor   1. A sensory cell or nerve ending that senses some change in the environment. 2. A sense organ containing sensory cells.

recessive   In genetics, a gene or trait that is concealed in the presence of a dominant gene or trait.

reflex   A subconscious response to a stimulus, such as the blinking of the eyes when an object flies toward them.

regenerate   To regrow a part of the body that is damaged or cut off.

relapsing fever   An infectious disease transmitted by lice and ticks, characterized by intermittent, high fevers.

retina   The concave, innermost layer of the eyeball. The retina contains receptors that sense light and transmit this information to the brain.

rickets   A weakening of the bones in children who are deficient in vitamin D or the parathyroid hormone. In rickets, there is less calcium available for the growth of bones, causing them to become bent and deformed.

rodent   A small, gnawing animal, such as a mouse, rat, or squirrel.

salamander   Any of the amphibians that resemble lizards but have smooth, moist skin. The larvae of salamanders are aquatic and have gills.

saliva   The fluid in the mouth. Saliva lubricates food, cleans the teeth, and starts the digestion of carbohydrates.

scalpel   A knife used for surgery and dissection.

scavenger   An animal that eats leftovers, often spoiled food.

scientific name   The Latin or Greek name given to different groups of plants or animals, for example, the genus and species of an insect.

sclera   The white, outer part of the eyeball.

scrotum   The pouch of skin and connective tissue that contains the testes.

sea anemone   A cylindrical animal with a ring of tentacles at its upper end that make it superficially resemble a flower. Its base is anchored in the rocks or coral of the sea.

seedling   A young plant.

segmentation   Alternating constriction (narrowing) and dilation (widening) of the wall of the intestine, dividing the intestinal contents into segments.

seine   A large, long net pulled through the water to catch fish and other animals.

semicircular canal   One of the three semicircular tubes of each inner ear that sense turning of the body.

semipermeable membrane   A membrane, such as that in dialysis tubing, that allows the passage of smaller particles, such as water molecules, but not the passage of larger particles, such as proteins.

sensor   A receptor cell or organ that senses a stimulus, such as light or heat.

sensory neuron   A nerve cell that transmits a sensory signal to the spinal cord or brain.

silkworm   The larva of a moth (especially the Asian moth, *Bombyx mori*) that spins large amounts of silk into its cocoon.

sinoatrial node (abbreviated SA node)   A group of modified muscle cells at the entrance to the heart that electrically discharge to generate the heartbeat.

skeletal muscle   Any of the rapidly contracting muscles, such as the biceps of the arm, that typically attach to and move bones.

slide A flat, rectangular piece of glass or plastic upon which specimens are placed for microscopic examination.

small intestine The winding tube in which food is digested and absorbed as it passes from the stomach to the large intestine.

smooth muscle Any of the slowly contracting muscles found in the digestive tract, respiratory tract, blood vessels, uterus, urinary bladder, and other internal structures.

soda lime A white, granular, alkaline substance that absorbs carbon dioxide and moisture.

sphincter A circular band of muscle that controls the opening and closing of a tube.

sphygmomanometer A device used to measure blood pressure.

spinal cord The cordlike mass of nerve cells in the vertebral canal of the backbone.

splice To unite by inserting, interweaving, or otherwise joining two items.

sponge Any of a group of highly porous, usually marine, invertebrate animals that draw food-bearing currents of water through their bodies.

stage The platform of a microscope upon which a specimen is placed for examination.

stereomicroscope A microscope with two eyepieces that allow objects to be seen in three dimensions.

stethoscope A device with which to listen to heartbeats or other sounds.

stimulus (plural stimuli) An agent, such as heat or electricity, that excites a sense organ, nerve, muscle, or gland.

stomach The curved, muscular pouch in which food is stored and partially digested before passing to the intestine.

stretch receptors Any sensory receptor that responds to stretching, such as a stretching of muscles.

substrate The support, such as soil, upon which a plant or animal lives.

suture 1. To sew together parts of the living body. 2. A thread used to sew the body. 3. A tight, immovable junction between two bones, as in the skull.

swarm To move about or emigrate in great numbers, as in a swarm of ants or honeybees.

sympathetic nerve One of the nerves originating in the thoracic and lumbar parts of the spinal cord and terminating in heart muscle, smooth muscle, and glands. Activation of sympathetic nerves causes the heart to beat faster and harder, respiratory passageways to dilate, and digestion to slow or stop.

talon The claw of an animal, such as an eagle.

telephoto lens A lens through which to take photographs of distant objects.

tendon A tough band of connective tissue that binds a muscle to a bone or to other tissue.

terrestrial Relating to land rather than water or air.

territorial marking Release of an odorous substance, such as urine, by an animal to show other animals where it normally lives.

testis (plural testes) A male reproductive organ in which sperm and male hormones, mainly testosterone, are produced.

testosterone   A male hormone produced by the testes. It promotes development of the male appearance and behavior.

thorax   The chest.

tincture of iodine   An orange solution of iodine in alcohol, sometimes used as an antiseptic.

tissue   A collection of similar cells that form a structural part of an organism. The primary tissues are nervous, epithelial, muscular, and connective.

tortoise   A turtle that lives on land.

tourniquet   A constricting band tied around an arm or leg to slow or stop bleeding.

toxin   A poisonous product of living cells.

trachea   The windpipe that connects the larynx to the bronchi.

tricuspid valve   *See atrioventricular valve.*

trowel   A scoop-shaped or flat-bladed tool used to dig holes for plants.

tuberculosis   An infectious disease that affects mainly the lungs.

typhus   An infectious disease characterized by high fever, rash, headache, and stupor or delirium.

urinary bladder   The pouch in which urine is stored.

urogenital   Pertaining to the urinary and reproductive systems.

uterine tube   One of the two uterine tubes that transport eggs from the ovary to the uterus. Also called *fallopian tube* or *oviduct.*

uterus (plural uteri)   The hollow, muscular pouch in which the embryo and fetus grow during pregnancy. Also called *womb.*

vagina   The distensible, muscular tube passing from the uterus to the exterior of the female body.

vein   Any of the thin-walled, branching vessels that carry blood from the venules to the heart.

ventral   On or toward the belly side of an animal.

ventricle   1. A thick-walled, muscular chamber of the heart. In mammals, the right ventricle pumps blood to the lungs, and the left ventricle pumps it to the rest of the body. 2. One of several cavities in the brain.

venule   Any of the narrow, thin-walled vessels that carry blood from the capillaries to the veins.

vertebra (plural vertebrae)   Any of the 33 bones that collectively form the vertebral column (backbone).

vestigial   Small and poorly developed. In fruit flies, vestigial wings are extremely short and wrinkled.

viable   Able to survive.

visual pigments   The light-sensitive pigments in the retina of the eye.

vole   A small, blunt-nosed rodent that resembles a mouse.

watch glass   A small, saucer-shaped piece of glass that resembles a watch crystal.

water soluble   Dissolving in water.

weevil   A beetle in which the head extends into a snout. The harmful larvae of weevils eat grain, fruit, and nuts.

wind instrument   A musical instrument through which sound is made by blowing.

woodchuck   A heavyset marmot of the northeastern United States and Canada. Also called *groundhog*.

xylol (formula C_8H_{10})   A transparent, oily liquid. Also called *xylene*.

Index